S0-FJQ-535

Library of
Davidson College

MOLECULAR BIOLOGY OF POLYOMAVIRUSES AND HERPESVIRUSES

MOLECULAR BIOLOGY OF POLYOMAVIRUSES AND HERPESVIRUSES

BERNARD V. PERBAL
School of Medicine
University of California, Los Angeles

HAWLEY K. LINKE
School of Medicine
Stanford University

GEORGE C. FAREED
Molecular Biology Institute
University of California, Los Angeles

A Wiley-Interscience Publication

JOHN WILEY AND SONS

New York   Chichester   Brisbane   Toronto   Singapore

576.6
P427m

Copyright © 1983 by John Wiley & Sons, Inc.

All rights reserved. Published simultaneously in Canada.

Reproduction or translation of any part of this work
beyond that permitted by Section 107 or 108 of the
1976 United States Copyright Act without the permission
of the copyright owner is unlawful. Requests for
permission or further information should be addressed to
the Permissions Department, John Wiley & Sons, Inc.

Library of Congress Cataloging in Publication Data:

Perbal, Bernard V.
  Molecular biology of polyomaviruses and herpesviruses.

  "A Wiley-Interscience publication."
  Bibliography: p.
  Includes index.
  1. Herpesviruses. 2. Polyoma virus. 3. Oncogenic
viruses. 4. Viral carcinogenesis. 5. Molecular biology.
I. Linke, Hawley K. (Hawley Karen) II. Fareed, George C.
(George Carr), 1944– . III. Title.

QR 400.P47 1983   576′.6484   82-23750
ISBN 0-471-05058-X

Printed in the United States of America

10 9 8 7 6 5 4 3 2 1

84-9265

Dedicated to

Annick
Tony
Edwige

PREFACE

Tremendous strides have been made in the past 10 years in understanding the molecular structures involved in the multiplicative cycles and processes of malignant transformation elicited by both DNA and RNA tumor viruses. We have selected members of two groups of DNA tumor viruses for presentation in this book: *Polyomavirus* and *Herpesvirus.* These viruses are at the same time among the simplest and the most complex of the animal and human viruses. With the development of modern biochemical and genetic engineering techniques for the elucidation of gene structure and function as well as protein identification and characterization, it has been possible to elucidate nearly completely the molecular biology of the polyomaviruses and to make great strides in the related work on such complex viruses as the herpesviruses. We hope that a presentation of the two different groups of viruses in this book will emphasize that particular point while providing updated information on the processes of gene organization and expression from these viruses. Let us keep in mind that much of our knowledge of the molecular mechanisms of gene expression in higher cells has come from studies of simpler genetic systems such as those represented in these viruses.

The principle cannot be enunciated too often: that anything a cell is seen to do in culture must be counted among its potentialities.

MARGARET R. MURRAY, 1977 (1)

vii

It has been nearly a century since Wilhelm Rous first used explants of the chick embryo medullary plate to study *in vitro* the mechanism underlying closure of the neural tube (2). In the early days of tissue culture, workers necessarily confined themselves to asking biochemical questions of a very general nature. The modern era began in the 1950s, with the exploitation (particularly by virologists and molecular biologists) of cell and tissue-culture techniques to answer specific problems (3). Tissue culture not only released the virologist from an unsavory dependence on developing chick and mouse embryos; it also facilitated the discovery of new pathogenic viruses as well as viruses with no known association or implication in disease processes.

REFERENCES

1. Fedoroff, S. and Hertz, L. (Eds.) (1977), *Cell, Tissue, and Organ Cultures in Neurobiology,* Academic Press, New York.
2. Parker, R. C. (Ed.) (1961), *Methods of Tissue Culture,* Paul B. Hoeber, New York.
3. Rothblat, G. H. and Cristofalo, V. J. (Eds.) (1972), *Growth, Nutrition, and Metabolism of Cells in Culture,* Academic Press, New York.

BERNARD V. PERBAL
HAWLEY K.LINKE
GEORGE C. FAREED

Los Angeles, California
August, 1983

CONTENTS

MOLECULAR BIOLOGY OF POLYOMAVIRUSES AND HERPESVIRUSES

POLYOMAVIRUSES

I. THE PAPOVAVIRIDAE

A. IDENTIFICATION AND GENERAL DESCRIPTION

In 1971, two new viruses were isolated from human tissues. For want of more descriptive terms, they were labeled with the initials of the patients from which they had been isolated, BK and JC. Both viruses proved to be related to the two already well-studied mammalian viruses: simian virus 40 (SV40) and polyomavirus (Py) (1,2). These four viruses now are considered to be individual species of the genus *Polyomavirus,* which belongs to the family Papoviridae along with the genus *Papillomavirus.* Besides their structural similarities, the genera share the Greek suffix *-oma,* used to form nouns denoting tumorous (3). All papovaviruses have been shown to cause tumors in either their natural host (papillomaviruses) or in different species from the species of origin of the virus (polyomaviruses) (4).

Although papillomaviruses were the first DNA viruses demonstrated to cause tumors (5), understanding of the molecular biology of these organisms has been severely impeded by their inability to grow under current tissue culture systems. The polyomaviruses provide a much more accessible experimental target. While on the trail of a leukemogenic RNA virus in 1953, Gross discovered a separate filterable agent that produced parotid tumors in mice (6). Because the parotid agent was shown to transform many different cell types *in vivo,* the term "polyoma virus" became universally accepted (4), eventually as one word.

SV40 was also discovered as a contaminant in another virus preparation. In nature, SV40 is found persistently associated with kidney cells of the rhesus monkey. These cells were used routinely to propagate the human adenovirus and poliovirus used in vaccines. Before this discovery thousands of patients were inoculated with live SV40 (7). The subsequent discovery by Eddy and her colleagues that SV40 causes tumors (8) inspired intensive in vestigations of both the virus and the vaccine recipients (4). Fortunately, no human disease has been linked unequivocally to an SV40 infection (although variants of SV40 have been isolated from certain human tissue). The original inspiration has led to innumerable contributions to our understanding of normal and abnormal cellular function (9).

All members of the *Polyomavirus* genus share many traits, not only in their gross physical properties (including specific regions of DNA sequence homology) but in their dynamics of growth as well.* They all contain approximately 3×10^6 daltons of double-strandard, circular DNA enclosed in a small (~45 nm diameter) icosohedral protein capsid (12). The DNA encodes five to seven known proteins. Three of these proteins are involved in capsid structure: VP1, VP2, and VP3 (4). In addition, at least four host-cell-encoded histones; H2A, H2B, H3, and H4, are associated with virion DNA and are responsible for a tertiary structure of the DNA similar to the cellular nucleosome (13,14) (also called minichromosomes).

B. BIOLOGICAL ACTIVITIES AND HOST RANGE

The polyoma virus—a DNA-containing virus—is characterized by a duality of actions: it produces neoplasias of various types in different species of rodents and causes cell degeneration in mouse embryo tissue cultures. . . . the results so far . . . suggest the existence of host-virus interaction with characteristics reminiscent of temperate bacteriophage.

Marguerite Vogt, 1960 (15)

* In addition to the four major *Polyomavirus* members there are the more recently discovered and poorly examined HD virus of the stump-tailed macaque (10) and the K virus of mice (11), which appear to be quite similar to BKV and JCV from initial superficial studies. They will not be discussed here.

In their classic report in 1960, Vogt and Dulbecco examined the two polar effects of Py virus infection. On one hand, an infection can cause cytocidal interaction leading to extensive viral synthesis and cell degeneration. On the other hand, a moderate interaction leads to the transformation of normal cells into neoplastic cells (15). These observations reflect both tissue culture and whole animal observations. Subsequently, it was observed that all polyomaviruses retain this duality of action to some degree in experimental tissue culture models. The full extent to which this holds true in the whole animal (particularly the human host) has yet to be determined. Cytocidal interaction is now most frequently referred to as lytic infection. The conversion to neoplastic characteristics is arrogantly described by those in the field simply as cellular transformation.

Each of the polyomaviruses has a separate and distinct host range—especially for lytic infection. SV40 grows well only in primary cells of African green monkey kidneys (AGMK), certain other monkey cells, and cell lines derived therefrom. By definition, it does not grow well in its persistently infected host–rhesus monkey kidney cells. It grew poorly, if at all, in human cells tested. SV40 has been shown to transform a variety of rodent cell cultures as well as certain human cells and readily causes the induction of tumors in newborn hamsters (especially gliomas and subcutaneous fibrosarcomas). When especially high titers of SV40 are used for intravenous inoculations, leukemias, lymphomas, osteosarcomas, and reticulum cell sarcomas develop. It does not induce tumors in the natural host nor, apparently, in humans (4,16,17).

Py virus multiplies in several cultures of mouse and other rodent cells, although it is endemic in most mouse populations. It has a broader range of oncogenicity than does SV40, transforming not only mouse cells but also cells of rats, rabbits, guinea pigs, dogs, cattle, monkeys, and man. Inoculations into animals (hamsters and mice are the most popular hosts) result in tumors in virtually all organs and tissues except the brain. Tumor formation in rabbits has also been reported (16,17).

BK virus, in contrast, has a very limited host range. Highest virus titers are obtained only when primary human cultures of brain, kidney, or endothelial origin (18–21) are used. Vero cells (monkey origin), while adequate for the original isolation of the virus, replicate BKV poorly after as few as three passages. In general, the BKV lytic cycle is slower than that of either SV40 or Py. Other monkey cultures such as primary AGMK, CV-1, or BSC-1, do not produce significant amounts of virus (18) and are frequently abortively transformed (22).* In culture, BKV has been shown to transform hamster kidney cells (23–25), rat embryo cells (26), rabbit cells (27,28), and several primary human cells including brain (29), kidney (20), and vascular endothelial cells (21). *In vivo,* BKV has been shown to be weakly oncogenic only in the newborn hamster (23,30–32), where it demonstrates a propensity toward cerebral tumors (33), and mastomy (9). It has been recovered, however, from the brain tumor of an immunosuppressed human (34). Reports of BKV DNA sequences in human tumors are conflicting and will be dealt with more fully below (35–37).

Finally, JCV appears to have the most limited host range of all. Its lytic cycle is also the slowest. Although an enormous number of cell lines have been investigated, productive infection occurs only in primary human fetal glial cells and in secondary human amnion cells (38), while human lung and kidney cells function only as very poor hosts (39). *In vitro,* it has been shown to transform only hamster (40) or human (41) glial cells, human endothelial cells (21), or human amnion cells (42). In contrast to BKV, JCV is highly oncogenic in newborn hamsters, primarily producing tumors associated with the brain (43,44). It has also been shown to induce tumors of the brain when inoculated into owl monkeys (45). In humans, JCV has been found repeatedly

---

* Abortive transformation implies partial advancement towards the transformed state but not maintenance.

associated with the degenerative disease, progressive multifocal leukoencephalopathy (PML) (46–48).

C. VIRUS REPLICATION AND EXPRESSION

1. Temporal Considerations

The *Polyomavirus* lytic infection cycle is initiated by adsorption and viropexis (pinocytosis) of virions into permissive cells. The virions lose their capsids, and the DNA may associate with host proteins and enter the nucleus (16). Because the virion DNA has a coding capacity for only a few proteins, it must depend on the host cell to supply most of the functions of viral replication, transcription, and translation. Unlike many other viral infections, infection with one of the polyomaviruses does not inhibit normal cellular functions. In fact, infection forces resting cells to reenter the cell cycle, and the entire complement of host DNA is subsequently replicated. Histone synthesis is also enhanced (following DNA replication). The synthesis of several known enzymes, the overall rate of protein synthesis, host RNA synthesis, and the transport of hexose sugars across the cell membrane are also enhanced prior to DNA synthesis (4).

The lytic cycle is divided temporally. Early functions are those that appear prior to the onset of viral DNA replication, beginning 12 to 15 hours after infection. Late functions are those expressed thereafter. In general, early expressions are nonstructural viral proteins required for induction of the host-cell functions mentioned above and the initiation of viral DNA replication; while late functions are primarily the virion structural proteins (4).

2. Functional Subdivision of the Viral Genome

Correspondingly, the viral genome is also divided into early and late components. Studies on temperature-sensitive (ts) and dele-

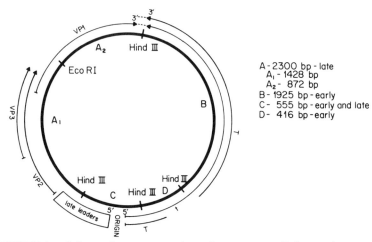

FIGURE 1. Schematic representation of a prototype *Polyomavirus* genome. Solid lines represent polypeptide coding regions; dashed lines equal full extent of 5′-capped and 3′-polyadenylated mature mRNA. Removal of RNA sequences by splicing events is indicated by *A*. General regions to which early mutations have been mapped are so noted.

tion (dl) mutants as well as analysis of mature messenger RNA (mRNA) species (4) and, ultimately, analysis of DNA sequences (49–55) have demonstrated that the *Polyomavirus* genome is virtually divided into functional halves. As shown in Fig. 1, all functions required during the early phase of infection are encoded by the right half of this composite genome, while the late (structural) functions are encoded by the left half.

Also indentified in Fig. 1 is the viral origin of DNA replication. Sequence analysis of SV40, Py, and BKV DNAs (49–55) indicates a high level of nucleotide conservation between viral origins which maintains the potential for extensive secondary structure. Early work which examined replicating structures by electron microscopy (56,57) or pulse-labeling (58) indicated that DNA replication proceeds at an equal rate in both directions from the single origin. Subsequently, a small portion of replicating Py molecules were also observed as rolling-circle replicative intermediates (59). Bidirectional replication terminates when the growing forks meet and the two daughter molecules are separated by nick and repair procedures (4).

3. Early Gene Expression

Production of virus-encoded RNA is the first gene expression observed following infection (60,61). It is the harbinger of the early viral proteins known at T antigens. Following the lead of those studying adenovirus-induced tumors SV40 (and Py), workers indentified a soluble antigen in both lytically infected and virus-transformed cells. It reacted with serum from virally induced tumor-bearing animals in a standard complement fixation assay (C'F) (62–65). The reactive property was distinct from virion capsid antigens and appeared much sooner after infection than did virion antigens. Because the C'F activity was dependent upon sera from tumor-bearing animals, it was referred to as tumor, or T antigens. These sera reacted specifically with the nuclei of both infected and virally transformed cells in an indirect immunofluorescence assay (66,67).

By 1977, a total of three distinct T antigen polypeptides encoded by the early region of Py was discovered, two of which were also noted in SV40 (these two are also encoded by early BKV and JCV DNA). The first detected was the large T antigen whose molecular weight has been estimated as ranging from 81,000 to 100,000 (68–70,236). It is located primarily in the nucleus of the cell (68), but some forms have also been reported to be associated with the plasma membrane (71,72).

The second tumor antigen found in all polyomaviruses has a molecular weight of 17,000 to 22,000 and has been referred to as small t antigen, or t antigen (70,73,74). Although it shares significant amino acid homology with the large T antigen, several experiments (including extensive genetic studies) have shown it to be a distinct protein, encoded by its own mRNA (49–55,70,73–86).

A third virally encoded tumor antigen has been identified only in Py-infected cells. Its molecular weight is estimated at 55,000 to 60,000 and accordingly it has been designated "middle T" (87). It also shares some, but not all, of its amino acid sequence with large T antigen and is encoded by its own mRNA (81–83,86). Searches for an analogous protein in other polyomavirus have not been successful, and examination of analogous DNA sequences suggests that no such protein can be encoded (49–55).

Various genetic evidence has indicated that one or another of the early viral antigens are required for specific steps in lytic infection and transformation (4,88–93). The early region of the DNA is divided into two complementation groups. The A group includes mutations that involve primarily the carboxyl half of the large T antigen, while the hr-t (Py) or dl 0.54-0.59 (SV40) groups involve mutations primarily in the region that encodes the unique portion of t antigen. (For Py, this also includes the region encoding the middle T antigen. The Py t and T genes have only recently been genetically separated (91). "A" mutations were first identified in ts mutants of SV40 and Py (4,88,92). The chemically induced mutants were isolated by their ability to form plaques at "permissive" temperatures (31–33°C) but not at "nonpermissive" temperatures (~40°C). Multiple analyses of these viruses demonstrated that a functional A gene was required to initiate (but not to complete) each new round of viral DNA replication (92,94). It is also required, probably only transiently, to induce transcription of late viral RNAs, and it appears to regulate the synthesis of its own mRNA (95,96). For our purposes, we consider the A gene synonymous with large T antigen.

The role of large T antigen in cellular transformation is not as consistent among the polyomaviruses as it appears to be for lytic infection. While it is clear that T is an absolute requirement for the initiation of SV40 transformation (88–93), Py transformation has been achieved with Py DNA molecules in which the coding region for large T has been interrupted by cleavage or insertion of foreign DNA (97–99). Functional T may also be required for the maintenance of the transformed state in all but the Py system (100).

Other properties associated with large T antigen include:

(a) It is phosphorylated (70,73) and poly-ADP ribosylated (101).

(b) It is primarily sequestered in the nucleus (66,67), although it has been reported in the plasma membrane (71,72).

(c) It can induce synthesis of cellular enzymes (4) and is responsible for increased host RNA synthesis (102).

(d) It is required for the integration of tandem repeats of Py DNA into the host genome, and it facilitates excision of viral genomes (103–106).

(e) It binds to DNA in general and has high affinity for a discrete region of the viral origin of DNA replication (107–109).

(f) It copurifies initially with both an ATPase activity and a protein kinase activity, although rigorous assignment of these activities to the T molecule has not been proved. In fact, only the ATPase activity is maintained when an adenovirus-SV40 T antigen chimeric protein is highly purified (110–112).

(g) It is most "active" as a tetramer, perhaps in complexes with host-encoded proteins (108–108a).

(h) Preparation of T antigen can induce tumor immunity, so-called TSTA activity (see below) (113–114).

(i) During both lytic infection and transformation it can be observed to bind to discrete RNA molecules (114a–114b).

Less well established are the properties associated with small t antigen:

(a) It is not phosphorylated (70,73).

(b) It does not bind DNA (73).

(c) It is not required for lytic infection of SV40 or BKV (54,75,78).

(d) It is not yet associated with any enzyme activity, although t/middle-T mutant virions contain histone proteins that are underacetylated in relation to wild-type virions (115).

(e) It may play a role in inducing resting cells to enter into the S phase of the cell cycle (116,117).

(f) It is not a requirement for maintenance of transformation in SV40-transformed cells. In Py-transformed cells, however, small t and/or middle T are an absolute requirement for maintenance of transformation. A BKV deletion mutant which deletes DNA encoding the carboxyl end of t (MMV strain) behaves no differently during lytic infection or transformation than do full-length genomes (54,88–90,97,117,118).

(g) It is found largely in the soluble fraction of cytoplasmic extracts (70).

(h) Both SV40 t and Py t (and/or middle T) appear to play a role in the disorganization of actin cable structure noted in transformed cells (119,120).

Middle T antigen of Py has yet another distinct set of characteristics:

(a) It is phosphorylated (70).

(b) It is frequently associated with the plasma membrane and is a prime candidate for the expression of TSTA activity (87).

(c) Along with small t, it is an absolute requirement for maintenance of Py transformation. Although Py-transformed cells frequently fail to retain the entire viral genome, the region encoding t and middle T is always found intact (97,118).

(d) In multimeric form it is associated with a protein kinase activity which apparently directs the unusual phosphorylation of one or more tyrosine residues within middle T antigen molecules (121–123a).

We are compelled to point out that the vast majority of information concerning the functions of the individual early gene products has been worked out exclusively in SV40 and Py. The early functions of BKV (and JCV, especially) have not been as thoroughly investigated. Rather, much of ongoing human *Polyomavirus* research is based on presumptions formulated by DNA sequence or virion structural homology information.

4. Late Gene Expression

Once viral DNA synthesis has commenced in earnest, structural (or late) proteins are detectable in the cell. Shortly thereafter, infectious virions are assembled and released (4). Although structural proteins are synthesized in the cytoplasm, they are rapidly transported and accumulated in the nucleus (124). Very late in infection, assembled virions accumulate in cytoplasmic vacuoles awaiting cell lysis for release (4).

SV40, Py, and BKV all contain three virus-encoded structural proteins of approximately the same size (125–127). Table I com-

TABLE I. Molecular Weight of Virion Structural Proteins

Protein	Py[a]	SV40[b]	BKV[b]
43,000	VP1	47,000	47,000
36,500	VP2	33,500	37,600
31,500	VP3	22,500	31,700

[a]Sata from Schaffhausen and Benjamin (126) and Kasamatsu and Nehorayan (127).
[b]Data from Eckhart et al (125).

pares the estimated molecular weights of VP1, VP2 and VP3 obtained by sodium dodecylsulfate (SDS) gel electrophoresis of purified virions.

VP1 is by far the major capsid protein, making up 70–80% of the total protein in SV40 and Py virions. A certain level of immunological cross-reactivity has been observed between SV40 and Py VP1 (128) and, to a slightly lesser extent, with the human polyomaviruses (128). Cross-species antigenic relationships between the minor structural proteins VP2 and VP3 have not been tested. Within species, VP2 and VP3 show extensive homology with each other but not with VP1 when analyzed by tryptic peptide mapping (4).

An apparently non-structural, basic protein of molecular weight 7900 has been identified accumulating late in SV40 lytic infection (128a–128b). Before its isolation a putative protein (referred to as the agnogene) was predicted from the DNA sequence corresponding to the 5' end of late mRNAs. While the function of this protein is unknown, its loss correlates with slower virus growth (128c). Although it has not been isolated, a similiar protein can be predicted from the DNA sequence of BKV.

5. Transcription and Posttranscriptional Modification of mRNA

At most, the entire *Polyomavirus* early region has the coding capacity for a single 90,000-dalton protein. And yet, the early genes seem to encode as much as 175,000 daltons worth of polypeptides. Although the explanation that these proteins shared precursor-product relationships was considered, the correct explanation resided in an understanding of early gene transcription and posttranscriptional modification. "Splicing" of mRNA was first reported with another DNA tumor virus, adenovirus 2 (129), and, shortly thereafter, in SV40 (130). Nearly all eucaryotic genes examined generate nascent transcripts larger in size than the mature mRNA found actively associated with the ribosomes. Prior to translation, the initial transcript can be shortened either by specific cleavage of nucleotides from the termini or by removal of internal segments and rejoining of cut ends—the mechanism referred to as splicing.

Prior to the onset of viral DNA synthesis, 10 to 100 mature mRNA molecules encoded by the early region accumulate as a single size class of SV40 early mRNA ∼19S; and two size classes, 18S and 19S, for Py, have been identified by buoyant density centrifugation. The late mRNAs for both viruses sediment at 16S, the predominant species, and 19S. These accumulate to 5,000 to 10,000 molecules per cell. All *Polyomavirus* mRNAs are both 3′-polyadenylated and 5′-⁷ meG capped (4).

Recent studies comparing conserved sequences between papovaviruses along with analyses of deletion mutants of SV40 (131,132) and of host range mutants of polyomaviruses (133) have defined a critical sequence near the origins for replication of SV40, Py, and BK. This 72-base pair, tandemly repeated sequence appears to have regulatory effects on both early and late gene expression and may be the essential binding site for RNA polymerase II.

All mRNAs isolated from *Polyomavirus*-infected or transformed cells are products of splicing. The approximate locations of splice junctions are noted in Figure 1. Estimates of splice junctions were obtained in most cases by R-loop mapping (129), restriction fragment hybridization (130), or by the application of the method first described by Berk and Sharp for SV40 transcripts (134). More precise splice junction assignments have been deduced, based on consensuslike sequences observed in viral DNA sequences (49–55).

The following is known about mRNA splicing in these viruses:

(a) **SV40**. Although the early mRNA was believed to be a single 19S species, there are actually two molecules that differ slightly in size, both of which are created by splicing mechanisms. Small t antigen is synthesized from the larger mRNA, and large T antigen is synthesized from the smaller (134–136). The mRNAs share the same 5' ends, though the T mRNA has an internal portion removed complementary to the SV40 DNA sequences between 0.54 and 0.59 map units. This is required because that region of the DNA has been shown to encode several translation termination signals in all three reading frames (49,50). The small t mRNA, however, does not exclude all of these termination signals (despite a small splice) and, therefore, translates only the smaller protein. This explains the tryptic peptide data which demonstrated that large T and small t contain five or six methionine-labeled peptides in common, and yet T contained at least an additional five unique peptides and t contained an additional two unique peptides (83,84).

The late mRNAs all include a noncontiguous leader sequence spliced onto the coding sequence of the message. The coding regions equivalent to the 19S RNA of VP2 and VP3 overlap one another and partially overlap the region encoding the 16S RNA of VP1; that is, the VP1 amino terminal region is encoded by the same DNA that encodes the terminating codons for VP2 and VP3. This overlap succeeds because the mRNA are translated in different reading frames. Although all the late leader sequences are similar, they are not identical. This difference may play a role in the quantitative regulation of the accumulation of the different species (130).

(b) Py. Polyomavirus early RNA makes maximum use of "frame shift" economy. In this case there are three separable early messages (86). All mRNAs contain identical sequences in their 5' segment and translate this region identically. As with SV40 early mRNA, identity ceases before the distal end of the t coding region (equivalent to the 0.54 to 0.59 region of SV40). Beyond this point, each of the three T antigens is translated from a messenger bearing a unique splice, the distal side of the splice positioned to translate each protein in a different reading frame (81,82).

The late Py messages are essentially the same as those for SV40.

(c) **BKV.** BK virus (like SV40) produces two species each of early and late mRNA. Evidence submitted thus far indicates the arrangement of splice segments is analogous to that described for SV40 (135,136).

(d) **JCV.** No analysis of JC virus-encoded mRNAs has been reported.

D. CELLULAR TRANSFORMATION

During transformation, viral expression is somewhat modified from that of lytic infection. In order to understand transformation, we will review viral-sepcific antigens not fully covered previously.

(a) **TSTA.** Tumor specific transplantation antigen (also abbreviated TrA) has been defined by its ability to confer immunity against tumor growth induced by injection of homologous infectious virus or virus-transformed cells (4). TSTA is observed in both lytic and transformed cells and has been shown to reside in the plasma membrane of the cell. The temporal appearance of TSTA is associated with T antigen and its activity is associated with an active A gene. Most SV40 investigators subscribe to the hypothesis that TSTA represents T antigens or a fragment thereof (distal to the 0.59 region of the DNA inserted in the plasma membrane (113,137–139). In Py-transformed cells, however, the likely TSTA candidate appears to be middle T antigen (87,139).

(b) **S.** S antigens include a poorly understood collection of antigenic activity associated with the cell surface generally, but not exclusively, during viral infection. In some cases, it appears they are cellularly encoded products induced to synthesis by the viral genome (4,140,141).

(c) **ANA.** Antinuclear immunofluorescent antibodies have

been observed by some investigators working with serum from tumor-bearing hamsters (23). They are observed in noninfected normal cells and have not been associated with any virally encoded protein.

(d) U. U antigen is observed as nuclear or perinuclear staining when antitumor serum is used. Although it is extremely more heat stable than T antigen, it is probably an alternate antigenic expression of the carboxyl region of the same polypeptide chain which expresses heat-labile T antigen (142,143).

(e) NVT. NVT is a newly coined acronym for nonviral tumor antigens, also called host-encoded antigens, Tau, p 53. These proteins have been identified by their appearance in immunoprecipitates of T/t antigens from virus-transformed cells (including transformants of SV40, Py, BKV, and JCV) (144,151). Although they are antigenically distinct from T/t antigens, they appear in preparations in which antitumor serum are used, primarily because they are complexed to either large T (144,145,147–150) or small t antigen 146,151). They have also been immunoprecipitated from chemical and spontaneous transformants as well as transformants induced by RNA tumor viruses (147,150) and normal cells (149) in which sera directed against purified NVT alone is used. In SV40-transformed (but not lytically infected) cells, this reaction conversely precipitates large T antigen (again as part of a complex) (150). Most NVTs have a molecular weight estimated around 55,000, although a 32,000 NVT has been reported associated with BKV t antigen (151). Peptide maps indicate little or no homology to virus-encoded antigens, although NVTs from a variety of cells show greater peptide homology to each other (148). Furthermore, the NVTs induced in hamster cells by several different polyomaviruses (SV40, BKV, and JCV) share identical tryptic peptide patterns (149). The function of NVTs in the transformed cells remains undetermined.

(f) Tiny t. A fourth T/t antigen has recently been reported by Spangler et al. (152) as a protein associated only with SV40-

transformed cells (not lytic infection) which migrates with an apparent molecular weight of 8000 following immunoprecipitation of SV80 cells (a transformed human fibroblast line). It shares eight or nine tryptic peptides with both large and small T antigens, does not bind DNA, is not phosphorylated, and is found only in the cytoplasm. It can be synthesized *in vitro* when two different SV80 mRNA species are used. Although it is found infrequently in other SV40-transformed cells, its function in transformation (if any) is unknown.

II. TUMORS AND THE TRANSFORMED CELL IN CULTURE

> I like to divide cancers into two categories . . . cancers that God made and cancers that virologists make . . . [I separate them because] I am not at all sure that God does what virologists like to do.
>
>   Albert B. Sabin, 1965 (153)

A. DEFINITIONS

Researchers frequently allude to "transformation" and its relationship to "neoplasia," "malignant" cell growth, or simply "cancer." At this point, it is useful to define the terms more fully and discuss the role (actual or potential) that DNA tumor viruses might play in the overall phenomenon (154). It is common to think of a cancer or tumor cell, either *in vitro* or *in vivo,* as a normal cell that no longer responds normally to signals associated with growth control. Of course, this does not imply totally random growth. Many tumors *in vivo* retain characteristics of their tissue of origin (17), while cells *in vitro* frequently display markers of their differentiation prior to transformation (e.g., specific hormones, pigment, etc.) (153,156).

The original tissue culture tumor models came from the transplantation of tumor cells from their original whole animal host to the *in vitro* environment (157). A major reservation in extrapolating back to the *in vivo* situation is that of selective pressure; that is, only those tumor cells that can successfully adapt to the new environment are observed. Nonetheless, several characteristics of tumor cells in tissue culture not observed with normal cells have been reported and linked to observed characteristics of tumors. We shall consider these in more detail below.

It was also noted that cells derived from normal (i.e., nontumorous) tissues could be induced to express many of these *in vitro* tumor cell characteristics. Such cells are referred to as transformed. The inducing stimuli included irradiation, chemicals (158) (generally mutagens), viruses (162) and time (4,160). The latter refers to induction in the absence of a known agent, so-called spontaneous transformation.

In vivo tumors fall into two categories. (a) Benign tumors that represent abnormal growth of specific cells, usually following a normal mitogenic stimulant (e.g., hormones). The tumors are not invasive (i.e., do not migrate into surrounding tissue) and, therefore, do not metastasize (i.e., form tumors at a distant site). Histologically these tumors are reasonably well organized. (b) Malignant tumors or cancers that develop into large masses of abnormal tissue and eventually spread by metastasis. They are histologically disorganized, and the cells within them demonstrate mitotic and karyotypic abnormalities (17).

B. COMPARISONS BETWEEN NATURALLY OCCURRING TUMORS AND TRANSFORMED CELL CULTURES

1. Cell Proliferation

In the adult animal, continued proliferation of cells is limited primarily to epithelial cells (e.g., of the skin, gut, and bronchii) and to stem cells (e.g., of thymus, bone marrow, and lymph nodes). Proliferation is controlled in balance with natural cell death. Other cells are capable of limited proliferation only in response to a specific mitogenic stimulus (e.g., liver cells following abnormal tissue damage or epithelial cells of the mammary glands). In the absence of such stimuli, tissues are generally devoid of mitotic structures (17). Still other cells, nerve cells and striated muscle, are irreversibly arrested in the G_0 phase of the cell cycle (161). A characteristic of all tumor cells is that they have escaped control of cell proliferation to some extent.

In tissue culture, proliferation of normal cells is also held in check. This has been attributed to sensation of cell density by cell-to-cell interaction (contact-mediated growth control) (162) and/or by accumulation of cell-produced inhibitory substances or depletion of essential substances in the extracellular environment (163,164). Whichever the control, proliferation of normal cells ceases at a given density, dependent upon the cell type and growth medium. Proliferation of tumor or transformed cells in culture is oblivious to the same signals. The requirements for transformed cell division are simply not as fastidious as for cell division of normal counterparts.

2. Positional Control

Normal cells are organized into tissues with specific spatial confinements and physical barriers. Cells of one tissue type do not penetrate into tissue of another type. This sort of invasion, however, is definitive of the cancer cell (17). It necessarily involves motility of the cancer cells and may also involve specific degradation of the surrounding normal tissue (160). While many normal cells in culture are mobile, such activity reverses or ceases when two cells come in contact, that is, contact inhibition of movement (165). Again, transformed cells have been shown to continue around or under another cell with which they have established contact (166).

3. Anchorage-Dependence

Another requirement for proliferation of normal cells may be the need to contact underlying structures. Proliferation of normal epithelial tissue occurs only among those cells in contact with the basal lamina. Cells that move into higher layers do not proliferate. Cells of skin carcinomas, however, do not require contact with basal lamina in order to divide. In this case, they have lost the contact or anchorage-dependence for proliferation. A totally analogous situation exists among tissue culture cells. While normal cells fail to undergo division when suspended in a semisolid medium, spontaneously transformed or virally transformed cells lose this requirement and can multiply in the suspended state (167).

4. Cell Structure

Another view of the substratum independence of transformed cells can be observed in their morphology. Normal cells tend to be flattened to the substrate and more firmly attached, while transformed cells are less adhesive and present a more rounded morphology (160). This is proportedly a reflection of alterations in both cells' surfaces and cytoskeletal components of transformed, but not normal, cells. Structural changes of the transformed cells observed *in vitro* might also contribute to the metastasizing cells' ability to withstand the mechanical rigors of its "fantastic voyage" through the afflicted host's blood vessels on the way to a distant site.

5. Multilayer Cultures

Changes in positional control, motility, and adhesion are all involved in the ability of transformed cells to overgrow the monolayer of normal cells and exist as multilayers in culture. In addition, transformed cells are "immortal" in culture, while most normal cells have a culture lifetime limited to a few passages (160).

6. Enzymology

There are several potential advantages for altered enzyme synthesis in cancer cells. (a) Enzyme-mediated modification has been suggested as contributing to adhesive and mobility properties associated with changes in the cell surface. (b) Proteolytic activity would be helpful in invasion of surrounding tissue. (c) Specific proteases may also be involved in formation of fibrin clots. Such clots could assist in trapping metastasizing cells at a new site for tumor growth (17).

Corroborative evidence for the above hypotheses has been demonstrated in a comparison of enzymatic capabilities of normal and transformed cultures, although *in vivo* cancer cell data are less supportive (160).

7. Tumorigenicity

The final and most compelling correlation between the *in vivo* and *in vitro* studies is that of tumorigenicity. The transplantability of a tumor from one host to another (syngeneic) host is a well-established principle (157). The ability of a cell transformed *in vitro* to induce tumors in an appropriate host is required for the most stringent definition of the transformed phenotype. [The *in vitro* growth criterion most closely associated with *in vivo* tumorigenicity is a demonstration of anchorage-independent growth (17).] Tumorigenicity is only suitably ascertained when histocompatibility differences between the transformed cells and potential host animal have been eliminated. When syngeneic animals are not available, investigators analyze tumor formation in the T-lymphocyte-deficient nude mouse. Only tumor cells (not normal cells) proliferate when injected into this host (160).

8. Disclaimer

The above correlations between tumors in the whole animal and transformed cells in culture are not absolute. They represent a repertoire of characteristics, some held in higher esteem than others, from which a transformed cell may choose. For each characteristic there are exceptions in tissue culture. They must, therefore, be considered indicative, not definitive, of the cancerous cell.

III.   VIRUSES AND THE TRANSFORMED CELL

> The tumor agent has been thrown down with the centrifuge, so too with many viruses.
>
> Peyton Rous, 1935 (168)

In the preceding section we described several specific characteristics of the transformed cell that are accessible to laboratory investigation. What role do polyomaviruses, and viruses in general, play in creating and maintaining these transformation-specific characteristics? Even before investigation had ascertained that viruses were living organisms (not merely chemical substances), it was clear that they caused tumors in certain animals. We now know that many RNA and DNA viruses are capable of creating tumors in most species of animals (amphibians to mammals). For the most part, however, we will confine ourselves to the very limited area of four viruses, Py, SV40, BKV, and JCV, and to the cells they have been shown to transform.

A.   STATE OF THE VIRAL DNA IN THE HOST CELL

Of all the transforming viruses known, polyomaviruses contain the smallest amount of genetic information. The mechanism by which this DNA effects its profound alterations on cells remains unclear. Sambrook et al. (169) first verified the hypothesis that the viral genome retained in SV40-transformed cells was physically connected to the host genome. This state of viral "integration" has been subsequently reported in all the SV40- and Py-transformed cells studied to date. Generally, 1 to 10 genome equivalents of SV40 DNA are present per diploid quantity of cellular DNA (4,170).

As mentioned above, the early work suggested analogies between viral transformation of higher eucaryotic organisms and the lysogenic bacteriophage. However, more extensive work showed that, unlike phage, SV40 and Py apparently display no specificity for the site at which they integrate. Furthermore, there is no preferred site at which the circular viral genome is opened to promote integration (171). Although the entire viral genome is frequently found integrated (often several copies thereof), it is not required in its entirety for either integration or transformation. Frequently, only a specific portion of the viral genome can be recovered as integrated sequences. This alone is sufficient to initiate and/or maintain the transformed state. Additionally, only specific viral proteins are expressed in transformed cells (4,170).

In those cases where the full viral genome is incorporated, it is usually impossible to find any evidence of infectious SV40 virus. Virus can frequently be induced, however, if the transformed cells are fused with appropriately permissive cells (4). Py-transformed cells, on the other hand, usually contain both integrated and nonintegrated viral genomes, although spontaneous release of Py virions is also a rare event (4).

Although JCV transformation has undergone very limited study, it appears similar to SV40. That is, 4 to 10 genome equivalents are integrated per diploid cell genome, probably in a tandem head-to-tail arrangement, at multiple sites. No free JCV sequences have been detected in transformants (172), although there has been one report of suspected episomal JCV in JCV-bacterial plasmid-transformed human amnion cells (57). JC virus can also be rescued from transformed cells by fusion with permissive cells (58).

In most instances, BKV also seems to adhere to the integration pattern of SV40 and Py. That is, BKV-transformed cells retain around five copies of integrated viral DNA in nonpermissive host systems (25,173). When permissive systems are transformed, free DNA can be found along with the integrated copies (20,174–176). An example of this interaction is observed when BKV interacts with and malignantly transforms human fetal brain cells without becoming integrated into the host genome (29). Since this initial report, similar BK virus–host relationships have been reported by others (177).

B.   ROLE OF INTEGRATION IN EFFECTING TRANSFORMATION

Does integration per se then play a significant role in cellular transformation? The fact that integration of viral genomes is associated with nearly every other tumor virus (both DNA and RNA viruses) is an important consideration. Two possible consequences of viral integration that might bear on transformation are immediately obvious. (a) The viral DNA could integrate within a cellular gene involved in normal growth control, thereby disrupting its production; or (b) integration could occur in regions important to control of normal gene expression (e.g., promotor regions). Although these physical perturbation theories were once favored, there is mounting evidence to indicate that expression of viral genomes is the major requirement for maintenance of transformation. Integration itself, as we will show, is definitely not a requirement for BKV transformation. Others have likewise shown this to be true for nonpapovaviruses as well (178,180). The phenomenon of viral DNA integration may be merely a convenient and available mechanism to ensure the maintenance of the viral DNA which encodes the crucial functions for transformation.

C.   ROLE OF VIRAL EXPRESSION IN EFFECTING TRANSFORMATION

There is little doubt that the required viral functions are products of the early DNA region in polyomaviruses. As we discussed above, only that region of DNA is required for transformation, and only that region is expressed in most virally transformed cells. Examination of appropriate mutants has established precisely which early functions of SV40 and Py are required for initiation versus maintenance of the transformed state.

In JCV-transformed cells, little work has been completed, though all transformed cells examined express intranuclear T antigen (40,42,43). Small T, though present in lytic infection, has not been investigated.

Most BKV transformants also express large T antigen by immunoflorescence. In general, they express small t antigen as observed by immunoprecipitation as well (19-21,23,24,26,30). A naturally occurring mutant of BKV (MMV strain) deletes the region of DNA encoding the carboxyl terminus of small t-antigen. This mutant is analogous to mutants of SV40 which deletes a similar region and are referred to as the 0.54–0.59 mutants. Unlike the SV40 mutants, however, MMV functions identically to wild type in the generation of tumors *in vivo* or cellular transformation *in vitro* (27). In one study, however, it has been shown that BKV transformation of human fetal brain cells does not require detectable expression of large T antigen.

It is generally assumed that polyomavirus transformation induces a widespread, pleiotrophic alteration of the host cell. Early viral antigen itself may act specifically in many ways. Here are the most obvious:

(a)   T antigen binds to DNA, particularly at the viral origin of replication. Similar sequences have been detected in repetitive monkey DNA and may also act as origins of replication (181). If T antigen binding induces viral DNA synthesis, it may induce "uncontrolled" cellular DNA synthesis.

(b)   T antigen is a negative regulator of early mRNA synthesis and a positive regulator of late mRNA synthesis. It may also regulate host transcription (184), possibly through DNA binding near the point of initiation of transcription.

(c)   T antigen has been associated with enzymatic activity (e.g., ATPase, kinase) (110–112). Such activities perpetrated on host proteins could influence cell growth. Several investigators have noted associations between T/t antigens and a myriad of alterations of host proteins (particularly those associated with the cytoskeleton (115,119,120,182,183) as well as activation of quiescent host genes (185).

(d)   Others have postulated from structural relationships (186) and biological activities (116,117,119) that T/t antigens function as replacements for normal growth factors or mitogens.

Tumor cells are notoriously afflicted with chromosome troubles.

W. H. Lewis, 1935 (187)

In addition, SV40 DNA, or its gene expression products, has been shown to be mutagenic for host cell DNA (188–192). The concept that cellular transformation is the result of a series of somatic mutations is frequently discussed. It is readily apparent that specific cellular mutations alone could induce the cascade of cellular alterations described above.

IV.   PERSISTENT INFECTION

Viruses are protected by the very cells that they infect, un-

less they kill these and thus expose themselves to neutralization by serum and to attack of of other sorts.

Peyton Rous, 1935 (168)

A. DESCRIPTIONS

Up to now we have focused our remarks on the virus-host relationship of transformation and acute lytic infection. We must
now address another capability of *Polyomavirus*, that is, pro-
longed lytic infection. We have arbitrarily separated lytic infections into types, based on their eventual outcome. Although our
definitions overlap with those of other authors, they do not necessarily coincide.

(a) In acute lytic infection *in vivo,* the virus enters the body
and proceeds to the appropriate tissue where multiplication takes place. The afflicted cells (though generally not
an entire tissue) are ultimately destroyed. Production and
release of viral antigens raises specific and nonspecific
host immunologic defenses and (except in the case of
especially virulent agents) the host rapidly eliminates the
infectious agent. In the vast majority of acute viral infections, there are no detectable clinical symptoms of disease
(193).

In tissue culture, acute lytic infection proceeds as
above except in the absence of most of the host's immune
responses. An exception in some cultures may be the ability of infected cells to induce interferon, thereby protect-
ing neighboring, and as yet uninfected, cells (194). In the
absence of an immune response, however, virtually all
permissive cells succumb during the course of lytic infection.

(b) Opposed to the rapid appearance and elimination of virus
in acute infection are several situations that can be com-
bined under the term *persistent infections.*

Slow virus infections display a long and variable incubation period between introduction into the host and appearance of detectable symptoms. Frequently the virus can be found replicating in tissue over an extended period of time (195,196). Two classic examples of slow viruses are visna, a sheep virus (197), and rabies, a virus found in virtually all warm-blooded animals (198).

In *steady-state* or *carrier virus infections,* the virus persists by continued infection of newly susceptible cells. The virus generally replicates slowly in relation to the rate at which uninfected cells multiply. The best example for our purposes is the apparent natural carrier culture of SV40 in rhesus monkey kidneys (193).

In *latent infections* the virus remains associated with the host cell in the absence of detectable cytopathology. The virus is frequently capable of reestablishing a lytic infection under appropriate circumstance. Herpes simplex (cold sores) and herpes zoster (shingles) viruses are perhaps the most widely known examples of latent infection. These viruses infect neurons in sensory ganglia without perturbing the cell. Occasionally, the virus undergoes detectable replication (without destroying the neuron) and virions migrate down the axon to the skin. There it manifests itself as disease by replicating in and killing epithelial cells (199). Other human diseases that are candidates for latent infection include subacute sclerosing panencephalitis (SSPE), associated with measles virus; and progressive multifocal leukoencephalopathy (PML), associated with JCV and SV40 (46,199,200).

B.   MECHANISMS

Several mechanisms have been postulated and, indeed, demonstrated to play a role in certain persistent infections.

1. Defective Interfering Particles (DIs)

Huang and Baltimore (201) were the first seriously to propose that defective interfering (DI) virus particles were responsible for persistent viral diseases. By their definition DIs are virus particles that (a) replicate only in the presence of helper virus; (b) contain only part of the viral genome (and, therefore, are defective) but make normal structural viral proteins; and (c) interfere specifically with the intracellular replication of nondefective homologous virus. Although DI particles from RNA viruses are the best understood, DIs have been reported for nearly all animal viruses (202), including association with DNA-persistent virus infections of SV40 (203,204) and herpes simplex (179). Examination of JCV recovered from PML patients however, indicates that neither DIs nor temperature sensitive mutants (see below) play a role in that disease (204a).

The basic theory of DI involvement in persistent infection is as follows: (a) Initially, normal virus replication leads to the generation of a few DI particles. (b) By definition, the DIs replicate preferentially to the standard "helper" particles. (c) When DI production is maximal, very few competent helper genomes remain to coinfect susceptible cells. (d) Since infection of DI particles only into a cell would fail to yield productive infection, levels of DIs accordingly fall off. (e) This leaves only rare wild-type "helper" particles to reinitiate the cycle of infection. This cycling effect is what maintains the persistent virus level (201).

2. Temperature Sensitivity

Dominance of replication of temperature-sensitive (ts) mutants over that of wild type genomes has also been reported in persistent infections (196). Persistent Aleutian virus disease of mink appears to be caused by a naturally and reproducibly selected ts mutant (205). Another approach to ts control of viral persistence can be observed in the wild in frogs. A herpeslike virus has been shown to induce the Lucké renal adenocarcinoma. Although tumor-bearing frogs can be found throughout the year, those captured in the warm-weather months contain virus particles. If frogs are captured during winter months or maintained in the laboratory at 4°C, the tumor tissue is virus-free (206).

3. Integration

The potential role of viral or proviral DNA integrated into the host genome in persistent infection is dependent upon the subsequent ability for "excision" and/or separate replication of the viral genome. This ability has been clearly demonstrated for the members of the *Polyomavirus* group as well as most others (4).

4. Mutation

Nonviable mutations of viral genes would lead to a failure to generate infectious progeny (in the absence of helper virus). It would not, however, necessarily interfere with the maintenance of the viral genome in an infected cell nor with production of viral proteins. Alterations in the host cell or other parameters of the environment might subsequently permit production of infectious virions (179,207,208). Although such mutations could include DI particles, "interference" per se in not a requirement of this mechanism. *In vivo,* viable virus mutations could aid in avoidance of host immunologic surveillance.

5. Host Cell Alterations

Mutations or more transient alterations in the host cell DNA might also generate resistance to complete viral replication and assembly (209).

6. Host Immune Responses

Finally, any viral mechanism that would diminish or eliminate susceptibility of virions or infected cells to the host immune response would play an important role in virus persistence *in vivo* (210,211). This would include the antigenic drift noted in many persistent infections (i.e., continual alterations in membrane-exposed viral antigens [196]).

V. CREATION OF AN *IN VITRO* MODEL TO STUDY HUMAN TRANSFORMING VIRUS AND PERSISTENT INFECTIONS

[The] exploration may lead nowhere in the search for a possible viral etiology of human malignancy, but it is something that needs to be done . . . because of the strong probability that human beings are not likely to be an exception to what has been observed in lower animals.

Albert B. Sabin, 1965 (153)

A. PREVIOUS MODEL SYSTEMS

Most frequently, the virus-host cell model systems exploited have been analogous, but not homologous, to virus-host systems that exist in nature. For example, SV40 has been studied almost exclusively in the more permissive African green monkey cells rather than in the rhesus monkey cells in which it is naturally found. Likewise, tumorigenic studies have been performed only in heterologous hosts. Moreover, the reductionist approach to molecular biology in general eliminates potentially important factors that may function in the whole animal host. Although these heterologous and/or reductionist models have yielded important information, all the observations made with the use of these models do not necessarily extrapolate to the interaction of polyomaviruses with their natural hosts *in situ.*

In particular, the human virus species have been studied exclusively in heterologous species or in cell culture. Therefore, the role of these viruses in human disease has been ascertained by analogy only. A later section will describe one model system designed to examine the human *Polyomavirus* BKV in an environment as close to nature as possible.

B. BKV AND JCV IN THE HUMAN HOST

Although it remains unclear exactly where BKV replicates in the human host, it is believed to be a human virus for three reasons:

(a) It has a very limited host range for lytic infection *in vitro,* strongly preferring human primary culture (brain, kidney, and other endothelial cultures) and, to a lesser extent, an occasional monkey cell line (4).

(b) It has been recovered only from human tissue (urine, kidney, brain) (1,34,213,214). (However, SV40 has also been recovered from human tissue on rare occasions (215,216). Additionally, BKV can be routinely recovered from immunosuppressed individuals, whether the immunosuppression is genetic (34), disease associated (e.g., advanced malignancy) (212), pharmacologically induced (i.e., immunosuppression of transplant recipients) (1,213,214), or a natural sequela to normal physiologic processes (e.g., pregnancy) (217,218). Although not rigorously proved, it is currently believed that this virus-shedding is the result of unveiling of an ongoing persistent infection and not *de novo* infection of the compromised host (4).

(c) Most compelling are the results of serological studies. In those human populations examined, BKV neutralizing antibodies are present in 50 to 90% of the adult population (219–223). Seroconversion occurs before the age of 14 years but has not been associated with any clinical symptoms (223).

By these three criteria, JCV is similar, including the recovery of infectious virions from both "normal" and immunosuppressed individuals (217,224–226). However, the appearance of JCV also firmly correlates with at least one human pathologic condition. JCV was first isolated from the diseased brain tissue of a patient with progressive multifocal leukoencephalopathy (PML) (2). Further investigations continue to link PML with strains of JCV (at least 39 separate human cases) (4,46–48,226) or, on rare occasions, with SV40 strain variants (235). JCV has also been demonstrated in spontaneous PML of a rhesus monkey (227). JCV has not been linked to any human malignancy when screened by a variety of immunologic techniques (228,229). It is interesting, however, that JCV can induce brain tumors in non-immunosuppressed owl monkeys (45). By so doing, it is the first instance of a human virus inducing such tumors in primates (4). Histologic examination of human PML tissue reveals oligodendrocytes which appear to be lytically infected, and astrocytes whose bizarre forms and unusual miotic figures are reminiscent of transformed cells *in vitro* (47).

The question of BKV association with human malignancy is also unresolved. Although serological screens of cancer patients demonstrate anti-BKV T or V antibodies in many patients, no group has been 100% positive for BKV or BKV-like antigens (215,216).

A more direct approach has been to analyze human tumors for the presence of BKV DNA sequences. In 1976, Fiori and diMayorca (35) examined DNA extracted from human tumors for BKV sequences by DNA:DNA reassociation kinetics with purified BKV DNA. They reported finding such sequences associated with some human tumor tissue (5/12) and human tumor-derived cell lines (3/4). They did not detect BKV DNA sequences in nontumor tissues. However, Wold et al. (36), using a different procedure to assay DNA:DNA hybridization, came to a contradictory conclusion. They analyzed 166 human tumors, seven human malignant cell lines, and 53 normal tissue samples—including some of those purported by Fiori and di-Mayorca to carry BKV sequences. Wold et al. reported that all human tissue hybridized with a BKV probe to a level 5% greater than did nonhuman tissue. There was, however, no specific enrichment for BKV sequences above this level in any tumor samples. They attributed the 5% hybridization to human sequences contained in their BKV probe. No further explanation for the discrepancy with the earlier paper was offered.

The Fiori and diMayorca group have subsequently replied (37). Using the sensitive southern hybridization technique, they have demonstrated both free and integrated BKV DNA sequences in tumor tissue and in normal tissue from a tumor-bearing individual. They examined 105 tumor and normal tissues and cell lines and identified BKV in 43 of the samples. (The significance, if any, of these findings in describing a BKV etiology for human malignancy may rely in part on the results obtained using BKV and human fetal brain cells discussed more fully below.)

To briefly review, unlike other polyomaviruses, BKV and JCV are of human origin. They are probably associated with persistent infections in nature which may begin in childhood and which last undetected throughout the lifetime of an immunocompetent individual. Although their preferred tissue for replication *in vivo* is not known (especially for BKV), both renal and neural tissues are likely candidates and have been demonstrated so *in vitro*. Like the other polyomaviruses, they are oncogenic in nonprimate hosts, and JCV is also oncogenic in a primate host.

C. PRELIMINARY STEPS TO DEVELOP AND DEFINE AN *IN VITRO* MODEL SYSTEM

For our study of human–*Polyomavirus* interaction, we chose to examine the *in vitro* relationship between primary human fetal brain cells and the more accessible of the two viruses, BKV. Although several strains of BKV have been subsequently isolated (212,230,231), our studies were performed with the prototype stock originally isolated by Gardner and her colleagues (1) and passaged *in vitro* most recently in HEK cells. This work has been performed in collaboration with Kenneth K. Takemoto and his research group at the National Institutes of Health (NIH), Bethesda, Maryland.

Human fetal brain cells were dispersed from 12- 16-week-old fetuses and cultivated on plastic dishes. Microscopic observations indicated that the culture contained primarily astrocytes and spongioblasts as reported for such cultures by Padgett et al. (2). Following infection with BKV (5.5×10^7 plaque-forming units per 75 cm^2 flask), most of the cells appeared to undergo productive infection as identified by viral cytopathic effect (CPE) by 4 days postinfection. By 8 days postinfection, the entire monolayer appeared to have succumbed to the effects of lysis and virion release. Media overlying the cells was changed every 5 days, and within 4 to 6 weeks foci of rapidly growing cells appeared. This is similar to the observations made by several groups when an SV40 infection is generated in diploid human cells (232–234). When these cells attained confluency, they were passaged and maintained as any cell line and assayed for biologic properties associated with a transformed phenotype.

Properties of BKV-infected human fetal brain cells (BKHFB) include (29):

(a)   The BKHFB cells are small, polygonal cells; the small size suggests that spongioblasts may have been the progenitors of these cells (235).

(b)   They display a population doubling time of 18 hours.

(c)   They form colonies when sparsely seeded on plastic or when seeded into soft agar. Values for colony-forming efficiency in soft agar vary between 0.5 and 5% for different experimants.

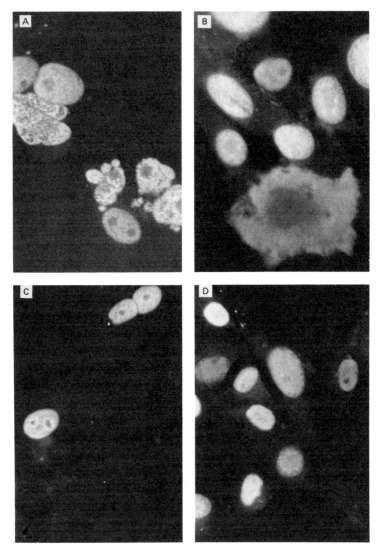

FIGURE 2. Human embryonic kidney cell monolayers infected 3 days before with BKV (prototype) stock virus (MOI≥ 1) were fixed in ice-cold acetone-methanol (1:1) and stained by indirect immunofluorescence. The primary antiserum reaction with the nuclei were BK-T (NIH) *(A)* or αBK-V (capsid) (NIH) *(B)*. Monolayers of BKV-infected human fetal brain cells were fixed and stained as above with αBK-T *(C)*, or αBK-V *(D)*.

(d) BKHFB cells will generate tumors when injected into nude mice. They are therefore tumorigenic.

(e) When assayed for the expression of viral antigens by immunofluorescence, BKHFB monolayers displayed between 5 and 30% positive nuclei. Although the percentage varied, at any given time the value was similar for both T and V antigen staining. In addition, infectious BK virions could be recovered continuously from the spent media and titered between 10^5 and 10^6 pfu per ml on human embryonic kidney cells (HEK); see Fig. 2. When assayed for hemagglutination, BKHFB spent media titered 1:10 to 1:160.

Two generalizations appeared to be true at this point. First, the primary human fetal brain cells had become malignantly transformed by association with BKV. Second, since most, but not all, of the cells were negative for detectable expression of viral antigens, a simple carrier state for persistence of BK virus may have been established.

In defense of the first principle, several observations should be noted. Human cells are generally very resistant to spontaneous transformations (235). Although more than 25 individual human fetal brains have been cultivated by Takemoto and his group, no spontaneous transformants have been observed (29). As with other diploid cell lines, the lifetime of human glial cells (the ultimate population of most embryonic brain cultures)is limited. In most cultures, the population begins to show signs of tissue culture senescence within 15 to 30 population doublings (around 2 to 3 weeks in exponentially growing cultures) (235). BKHFB cells had already been maintained for many months by passaging every 4 to 5 days at split ratios of 1:5 and showed no signs indicative of senescence.

The first principle (that BKV induced the transformation) was further supported by experiments designed to analyze the carrier state of the persistent viral infection, that is, the second apparent

Library of
Davidson College

principle. The original BKHFB cultures were cloned in medium containing 0.3% anti-BKV neutralizing antiserum (100 hemagglutination inhibition units/ml) by seeding sparsely (15–20 colonies per dish). Well-isolated colonies developed in 2 weeks and were individually propagated for further study. Morphologically, all clones appeared identical to the parental cells. When assayed for viral antigen by immunofluorescence, all clones were negative for both T and V staining. If the neutralizing antiserum was removed from the medium, however, cells began displaying immunofluorescence-positive T and V antigens with 3 to 4 weeks, and hemagglutination titers were again detected in the spent media.

This chain of events eliminated the possibility that the virus was maintained by a simple carrier culture in BKHFB monolayers since all virus had been removed from the extra cellular and intracellular environment by the antiserum. It did, however, underscore that even in the absence of infectious virus or detectable viral antigens, the BKV genome remained permanently associated with the transformed cell.

The parental BKHFB cell line was cloned in the presence of virus-neutralizing antiserum. Although treatment with antiserum eliminated all detectable virus expression, when the treatment was discounted, all isolated clones could be divided into two groups. The first group, like the parental culture, began to express viral antigens and virus and were designated "incurable". The second group, "cured", failed to demonstrate detectable viral expression.

Molecular analyses of BKHFB parental cells and all clonal cell lines have demonstrated the following. (a) All BKHFB cells retain the BKV genome as 1 to 10 copies per cell of nonintegrated, wild type, superhelically coiled DNA. (b) Replication of the viral genome and maintenance of the transformed phenotype can occur in the absence of detectable viral antigen. (c) BKHFB cells consistently express a unique host-encoded protein that may be associated with the transformed state. (d) Treatment of virus-shedding cultures with rabbit BKV-neutralizing antiserum (directed primarily against virion structural proteins) results in the suppression of all viral protein synthesis. (e) This suppression probably acts on the cell prior to the maturation of stable, virus encoded mRNA. (f) Under special circumstance, all viral DNA can be excluded from the host cell without significantly affecting phenotypic expression of malignant transformation.

Some of the observations outlined above on the interaction between a human polyomavirus and cells of its natural host are in contrast with observations of others examining (a) nonhuman, but highly homologous, polyomaviruses, or (b) human polyomaviruses in heterologous hosts. Although human malignancies have been examined by others for tumor virus gene expression and for persistence of the viral genome, the results have not demonstrated any firm correlations. These observations would indicate, however, that human *Polyomavirus*-induced malignancy would not necessarily be disclosed by application of current experimental approaches.

The ability to manipulate *Polyomavirus* expression in cells of the natural host by the addition of neutralizing antiserum may serve as a model for studies of persistent infection *in vivo*. Superficially, this "transmembrane" modulation appears similar to that reported by Fujinami and Oldstone for measles virus expression (210). They noted that antibody treatment of HeLa cells acutely infected with measles decreased the amount of both a virion structural protein expressed on the cell surface and a structural polypeptide found inside the cell. They concluded that antibody directed against a cell-surface antigen could not only strip off surface viral determinants but could also interfere with viral polypeptides not associated with the plasma membrane. They proposed that a similar function *in vivo* might contribute to the measles-like virus persistence of subacute sclerosing panencephalitis (SSPE).

Unlike measles, BKV is not a "budding" virus and, therefore, does not routinely express virion structural proteins on the infected cell surface. Although this should be considered when relating the measles phenomenon to BKHFB cells, one should also remember that the polyomaviruses do express surface antigens (e.g., TSTA) of a nonstructural nature during both lytic infection and transformation. These nonstructural proteins have been associated with regulatory functions. If, as we suspect, plasma-membrane-bound T antigen is frequently the only virus-encoded protein expressed in BKHFB cells, it would be available to facilitate neutralizing antiserum modulation of proteins within the cell (237). (Anti-T activity is frequently observed in neutralizing serum, even though the major activity is directed against VP1.)

A possible mechanism for antiserum modulation may be separate from the neutralizing antibody molecules found therein. Antiviral sera frequently carry another "transmembrane" modulator: interferon. Interferon has been reported to act at the level of inhibition of either viral-specific transcription or translation, depending on the system analyzed (238). Several investigators have specifically documented the effect of exogenously supplied interferon preparations on expression of early SV40 gene products. Kingsman and her colleagues have demonstrated that interferon abolishes synthesis of 94K, 62K, and 19K immunoprecipitable proteins of infected cells (239) by reducing synthesis and nuclear and cytoplasmic accumulation of early mRNA (240). Furthermore, Mozes and Defendi (241) reported that interferon-activated inhibition may be directed preferentially at "free" viral genomes rather than at integrated molecules.

VI. MECHANISM OF TRANSFORMATION

Many chronic irritants of diverse character give rise to cancer. . . .Has the virus a nearer relationship to malignancy than those irritants whose role is ended once the cancer has begun?

Peyton Rous, 1935 (168)

As discussed previously, one or all of the early *Polyomavirus* gene products are required for the separate steps of (a) initiation and (b) maintenance of the transformed state induced by Py or SV40. If we assume that our inability to detect early viral gene products is caused by failure of cells to make these products, how then has BKV induced the transformed state?

In vivo, tumor cell induction is purported to involve at least two steps, the first termed initiation and the second promotion. The classic example of this two-step process was reported in the 1940s when researchers observed that application of croton oil to the skin of mice enhanced the tumorigenic effect of chemical carcinogens. Croton oil alone (the active ingredients for promotion are phorbol esters such as 12-O-tetradecanoylphorbol 13-acetate-TPA), however, had no carcinogenic activity. Since that original work, many similar observations have been reported for *in vivo* and *in vitro* malignant transformations (242).

Recently, Seif and Martin (together and separately) have vigorously applied this concept to tumor-virus induced transformation, especially transformations induced by polyomaviruses. Together (116), they propose that the less-than-consistent data from several laboratories concerning the role of small t antigens in transformation is best explained by viewing this protein as the tumor promotor (acting like TPA, for example). Seif goes on to point out that, *in vivo,* most initiators are mutagens which may induce dominant or recessive mutations (245). Others have proposed that promotors act by then eliminating the dominant, normal allele to expose recessive mutations (243,244). Following this line of reasoning, the mutagenic activity attributed to SV40 may account for at least a portion of its transforming ability. Seif further proposes that mutagenesis might be a function of repeated attempts to integrate, attempts that fail and leave behind broken DNA strands.

By analogy, BKV DNA may be mutagenic in HFB cells. Token advances toward integration, thwarted by insufficient or nonexistent T antigen, might generate appropriate mutations. The longer the viral DNA association, the more likely the mutations become significant to growth-control responses. If one adheres to this philosophy of viral transformation, the viral genome becomes ultimately dispensable, as in the case of the BKHFB clone which carries no detectable viral DNA.

However, SV40, Py, adenovirus, and the other tumorigenic viruses are all integrated in the transformed cell. Integration may occur simply because it is permitted to occur and has no specific value for virus propagation, that is, there is an absence of host factors that would prevent integration. Even though transformation per se may be permanently fixed within the cell (as in tsA transformants that lose their temperature-dependence), the viral genome is now permanently ensconced. Elimination of an unnecessary and unintegrated portion of viral DNA may be a more likely occurrence than the elimination of integrated DNA.

A.   MALIGNANT TRANSFORMATION AND BKV

> The tumor problem has withstood the most corrosive reasoning. Yet since what one thinks determines what one does in research, it is well to think something. And it may prove worthwhile to think that one or more tumors of unknown cause are due to viruses.
>
>   Peyton Rous, 1935 (168)

Although the mechanisms are unclear, immunosuppression of the sort that leads to *Polyomavirus* shedding also leads to increased risk of malignancies (246,247). Aside from that general phenomenon, tumor production in brain tissue must be considered as an exceptional case. Since most neural cells in the adult brain are arrested in G_0, there is little capacity to regenerate, and cellular components are replaced at a very slow rate. Cell cycle factors appear to play a role in viral-induced transformation, and it may not be permissible to extrapolate from rapidly growing cell culture models to tumorigenicity of brain tissue. Extrapolation is more acceptable in describing *in vivo* transformation of other cell types.

We previously discussed the conflicting reports relating to the presence of BKV genomes in human tumor tissues. The most recent southern-blot data are most compelling and most consistent with observations in nature. That is, BKV sequences can be found associated with both normal and tumor tissue. The consistently high level of BKV (and JCV) neutralizing antibody observed in humans of all ages (and followed over several years) is most likely maintained by persistent infection. Examination of the sera from renal transplant patients, before and after the treatment which precedes BKV shed, indicates significant levels of antibody prior to immunosuppresion (4).

If the BKV genome persists in humans, it seems unlikely that examination of hundreds of human tissue samples would fail to detect BKV sequences in at least a few samples. We suggest that the 5% level of hybridization of BKV DNA to all human tissues observed by Wold et al. may be due, in part, to persistent viral genomes.

We cannot determine if BKV is involved in human tumorigenesis, but our work does bear on the interpretation of the studies of others. First, BKV is competent to cause malignant transformation of more than one human cell type in culture. This has been shown by others using kidney and endothelial cultures and, in this report, using brain cultures. Second, as demonstrated in several studies, large copy numbers of viral DNA are not required for transformation, nor is integration. Third, T antigen or other early or late gene product accumulation need not be consistently detected for a cell to appear fully transformed. Fourth, we have at least one example in which a "hit and run" type of viral-induced transformation may have occurred in human cultures which no longer retain any evidence of the viral genome.

B. IMPLICATIONS FOR HUMAN CANCER

In toto, these findings make it impossible to eliminate a BKV etiology for human tumors by currently applied techniques. The failure of electron microscopy studies to detect assembled virions in tumor cells is not relevant to transformation with any *Polyomavirus.* The inability to observe uniformly early human *Polyomavirus* gene products in human tumors is not contradictory to such viral etiology as we have shown in the BKHFB cultures. Finally, the random association of human *Polyomavirus* DNA with human tumors and normal tissue fulfills a natural expectation based on epidemiological studies. It, too, may be irrelevant to tumorigenesis. As long as BKV remains associated with human tissue, however, the potential for malignant transformation cannot be ignored.

It may be that constant low levels of *Polyomavirus* expression *in vivo* constantly inspires the host immunologic surveillance mechanisms. In the long run, such surveillance could remove host cells that may be in the early stages of viral-induced transformation. Any alteration of host or virus that permanently or temporarily circumvents this initial surveillance may ultimately contribute to tumor formation.

REFERENCES

1. Gardner, S.D., Field, A.M., Coleman, D.V., and Hulme, B. (1971), *Lancet* **i**:1253.

2. Padgett, B.L., Walker, D.L., ZuRhein, G.M., Eckroade, R.J., and Dessel, B.H., (1971), *Lancet* i:1257.

3. Fenner, F. (1976) *Intervirology* 7:20.

4. Tooze, J. (Ed.) (1980) *Molecular Biology of Tumor Viruses, Part 2, DNA Tumor Viruses* (2nd Edition), Cold Spring Harbor Laboratory, Cold Spring Harbor, New York.

5. Shope, R.E. (1932), *J. Exp. Med.* **56**:803.

6. Gross, L. (1953), *Proc. Soc. Exp. Biol. Med.* **83**:414.

7. Sweet, B.H. and Hilleman, M.R. (1960), *Proc. Soc. Exp. Biol. Med.* **105**:420.

8. Eddy, B.E., Borman, G.S., Grubbs, G.E., and Young, R.D. (1962), *Virology* **17**:65.

9. Khoury, G. Lai, C.-J., Brown, D., Israel, M., and Howley, P. (1977), in *Origins of Human Cancer, Book B: Mechanisms of Carcinogenesis.* H.H. Hiatt, J.D. Watson, and J.A. Winston, (Eds.), Cold Spring Harbor Laboratory, Cold Spring Harbor, New York.

10. Steffen, M., Krieg, P., Pernfuss, M., Sauer, E., Eisinger, V., and Sauer, G. (1980), *J. Virol.* **35**:865.

11. Law, M.-F., Takemoto, K.K., and Howley, P.M. (1979), *J. Virol.* **30**:90.

12. Fenner, F., McAuslan, B.R., Mims, C.A., Sambrook, J., and White, D.O. (Eds.) (1974), *The Biology of Animal Viruses* (2nd Edition), Academic Press, New York.

13. Fareed, G.C. and Davoli, D. (1977) *Ann. Rev. Biochem.* **46**:471.

14. Meneguzzi, G., Pignatti, P.F., Barbanti-Brodano, G., and Milanesi, G. (1978), *Proc. Natl. Acad. Sci. USA* **75**:1126.

15. Vogt, M. and Dulbecco, R. (1960), *Proc. Natl. Acad. Sci. USA* **46**:365.

16. Andrewes, C., Pereira, H.G., and Wildy, P. (Eds.) (1978), *Viruses of Vertebrates* (4th Edition), Baillière Tindall, London.

17. Weil, R. (1978), *Biochim Biophys Acta* **56**:301.

18. Takemoto, K.K. and Mullarkey, M.F. (1973), *J. Virol.* **12**:625.

19. Rundell, K., Tegtmeyer, P., Wright, P.J., and diMayorca, G. (1977), *Virology* **82**:206.

20. Purchio, A.F. and Fareed, G.C. (1979), *J. Virol.* **29**:763.

21. Fareed, G.C., Takemoto, K.K., and Gimborne, M.A., Jr. (1978), *Microbiology,* 1978, 427.

22. Mason, D.A., Jr. and Takemoto, K.K. (1976), *J. Virol.* **17**:1060.

23. Takemoto, K.K. and Martin, M.A. (1976), *J. Virol.* **17**:247.

24. Seehafer, J., Salmi, A., and Colter, J.S. (1977), *Virology* **77**:356.

25. Olive, D.M., Lampert, M., and Major, E.O. (1980), *Virology* **103**:1.

26. Farrell, M.P., Mäntyjärvi, R.A. and Pagano, J.S. (1978), *J. Virol.* **25**:871.

27. Mason, D.H. and Takemoto, K.K. (1977), *Int. J. Cancer* **19**:391.

28. Portolani, M., Barbanti-Brodano, G., and LaPlaca, M. (1978), *J. Gen. Virol.* **38**:369.

29. Takemoto, K.K., Linke, H., Miyamura, T., and Fareed, G.C. (1979), *J. Virol.* **29**:1177.

30. Van der Noordaa, J. (1976), *J. Gen. Virol.* **30**:371.

31. Shah, K.V., Daniel, R.W., and Strandber, J.D. (1975), *J. Natl. Cancer Inst.* **54**:945.

32. Uchida, S., Wantanabe, S., Aizawa, T., Furuno, A., and Muto, T. (1979), *J. Natl. Cancer Inst.* **63**:119.

33. Coralinni, A., Altavilla, G., Cecchetti, M.G., Fabris, G., Grossi, M.P., Balboni, P.G., Lanza, G., and Barbanti-Brodano, G. (1978), *J. Natl. Cancer Inst.* **61**:875.

34. Takemoto, K.K., Rabson, A.S., Mullarkey, M.F., Blaese, R.M., Garon, C.F., and Nelson, D. (1974), *J. Natl. Cancer Inst.* **53**:1205.

35. Fiori, M. and diMayorca, G. (1976), *Proc. Natl. Acad. Sci. USA* **73**:4662.

36. Wold, W.S.M., Mackey, J.K., Brackmann, K.H., Takemori, N., Rigden, P., and Green, M. (1978), *Proc. Natl. Acad. Sci. USA* **75**:454.

37. Pater, M.M., Pater, A., Fiori, M., Slota, J., and diMayorca, G. (1980), in *Viruses in Naturally Occurring Cancers* (M. Essex, G. Todaro, and H. zurHausen Eds.), p. 329, Cold Spring Harbor Laboratory, Cold Spring Harbor, New York.

38. Takemoto, K.K., Howley, P.M., Miyamura, T. (1979), *J. Virol.* **30**:384.

39. Frisque, R.J., Martin, J.D., Padgett, B.L., and Walker, D.L. (1979), *J. Virol.* **32**:476.

40. Frisque, R.J., Rifkin, D.B., and Walker, D.L. (1980), *J. Virol.***35**:265.

41. Walker, D.L. and Padgett, B.L. (1978), in *Microbiology* (D.

Schlessinger, Ed., p. 432, American Society for Microbiology, Washington, D.C.

42. Howley, P.M., Rentier-Delrue, F., Heilman, C.A., Law, M.-F., Chowdhury, K., Israel, M.A., and Takemoto, K.K. (1980), *J. Virol.* **36**:878.

43. Walker, D.L., Padgett, B.L., ZuRhein, G.M., Albert, A.E., and Marsh, R.F. (1973), *Science* **181**:674.

44. Padgett, B.L., Walker, D.L., ZuRhein, G.M., and Varakis, J.N. (1977), *Cancer Res.* **37**:718.

45. London, W.T., Houff, S.A., Madden, D.L., Fuccillo, D.A., Gravell, M., Wallen, W.C., Palmer, A.E., Sever, J.L., Padgett, B.L., Walker, D.L., ZuRhein, G.M., and Ohashi, T. (1978), *Science* **201**:1246.

46. Narayan, O., Penny, J.B., Jr., Johnson, R.T., Herndon, R.M., and Weiner, L. (1973), *New Engl. J. Med.* **289**:1278.

47. Padgett, B.L., Walker, D.L., ZuRhein, G.M., Hodach, A.E., and Chou, S.M. (1976), *J. Infect. Dis.* **133**:686.

48. ZuRhein, G.M. and Chou, S.-M. (1965), *Sciences (N.Y.)* **148**:1477.

49. Fiers, W., Contreras, R., Hageman, G., Rogiers, R., Van de Voorde, A., Van Herreweghe, J., Volckaert, G., and Ysebaert, M. (1978), *Nature* **273**:113.

50. Reddy, V.B., Thimmappaya, B., Dhar, R., Subramanian, K.N., Zain, B.S., Pan, J., Ghosh, P.K., Celma, M.L., and Weissman, S. M. (1978), *Science* **200**:494.

51. Friedman, T., Esty, A., LaPorte, P., and Deininger, P. (1979), *Cell* **17**:715.

52. Soeda, E., Arrand, J.R., Smolar, N., and Griffin, B.E. (1979), *Cell* **17**:357.

53. Arrand, J.R., Soeda, E., Walsh, J.E., Smolar, N., and Griffin, B.E. (1980), *J. Virol.* **33**:606.

54. Seif, I., Khoury, G., and Dhar, R. (1979), *Cell* **18**:963.

55. Yang, R.C.A., Young, A., and Wu, R. (1980), *J. Virol.* **34**:416.

56. Fareed, G.C., Garon, C.F., and Salzman, N.P. (1972), *J. Virol.* **10**:484.

57. Crawford, L.V., Syrett, C., and Wilde, A. (1973), *J. Gen. Virol.* **21**:515.

58. Danna, K.J. and Nathans, D. (1972), *Proc. Natl. Acad. Sci. USA* **69**:3097.

59. Bjursell, G. (1978), *J. Virol.* **26**:136.

60. Benjamin, T.L. (1966), *J. Mol. Biol.* **16**:359.

61. Sambrook, J., Sharp, P.A., and Keller, W. (1972), *J. Mol. Biol.* **70**:57.

62. Black, P.H., Rowe, W.P., Turner, H.C., and Huebner, R.J. (1963), *Proc. Natl. Acad. Sci. USA* **50**:1148.

63. Sabin, A.B. and Koch, M.A. (1964), *Proc. Natl. Acad. Sci. USA* **52**:1131.

64. Hoggan, M.D., Rowe, W.P., Black, P.H., and Huebner, R.J. (1965), *Proc. Natl. Acad. Sci. USA* **53**:12.

65. Habel, K. (1965), *Virology* **25**:55.

66. Pope, J.H. and Rowe, W.P. (1964), *J. Exp. Med.* **120**:121.

67. Rapp, F., Butel, J.S., and Melnick, J.L. (1964), *Proc. Soc. Exp. Biol. Med.* **116**:1131.

68. Tegtmeyer, P. (1975), *Cold Spring Harbor Symp. Quant. Biol.* **39**:9.

69. Ito, Y., Spurr, N., and Dulbecco, R. (1977), *Proc. Natl. Acad. Sci. USA* **74**:1259.

70. Schaffhausen, B.S., Silver, J.E., and Benjamin, T.L. (1978), *Proc. Natl. Acad. Sci. USA* **75**:79.

71. Soule, H.R. and Butel, J.S. (1979), *J. Virol.* **30**:523.

72. Deppert, W., Hanke, K., and Henning R. (1980), *J. Virol* **35**:505.

73. Prives, C., Gilboa, E., Revel, M., and Winocour, E. (1977), *Proc. Natl. Acad. Sci. USA* **74**:457.

74. Simmons, D.T. and Martin, M.A. (1978), *Proc. Natl. Acad. Sci. USA* **75**:1131.

75. Staneloni, R.J., Fluck, M.M., and Benjamin, T.L. (1977), *Virology* **77**:598.

76. Eckhart, W. (1977), *Virology* **77**:589.

77. Fluck, M., Staneloni, R.J., and Benjamin, T.L. (1977), *Virology* **77**:610.

78. Shenk, T.E., Karbon, J., and Berg, P. (1976), *J. Virol.* **18**:664.

79. Crawford, L.V., Cole, C.N., Smith, A.E., Paucha, E., Tegtmeyer, P., Rundell, K., and Berg, P. (1978), *Proc. Natl. Acad. Sci. USA* **73**:117.

80. Sleigh, M.J., Topp, W.C., Hanich, R., and Sambrook, J.F. (1978), *Cell* **14**:79.

81. Hutchinson, M.A., Hunter, T., and Eckhart, W. (1978), *Cell* **15**:65.

82. Smart, J.E. and Ito, Y. (1978), *Cell* **15**:1427.

83. Simmons, D.T., Chang, C., and Martin, M.A. (1979), *J. Virol.* **29**:881.

84. Linke, H.K., Hunter, T., and Walter, G. (1979), *J. Virol.* **29**:390.

85. Simmons, D.T., Takemoto, K.K., and Martin, M.A. (1977), *J. Virol.* **24**:319.

86. Hunter, T., Hutchinson, M.A., and Eckhart, W. (1978), *Proc. Natl. Acad. Sci. USA* **75**:5917.

87. Ito, Y., Brocklehurst, J.R., and Dulbecco, R. (1977), *Proc. Natl. Acad. Sci. USA* **74**:4666.

88. Fried, M. (1965), *Proc. Natl. Acad. Sci. USA* **53**:486.

89. Eckhart, W. (1969), *Virology* **38**:120.

90. diMayorca, G., Callender, J., Marin, G., and Girodano, R. (1969), *Virology* **38**:126.

91. Ito, Y., Spurr, N., and Griffin, B.E. (1980), *J. Virol* **35**:219.

92. Tegtmeyer, P. (1972), *J. Virol.* **10**:591.

93. Cowan, K., Tegtmeyer, P., and Anthony, D.D. (1973), *Proc. Natl. Acad. Sci. USA* **70**:1927.

94. Francke, B. and Eckhart, W. (1973), *Virology* **55**:127.

95. Tegtmeyer, P., Schwartz, M., Collins, J.K., and Rundell, K. (1975), *J. Virol.* **16**:168.

96. Reed, S.I., Stark, G.R., and Alwine, J.C. (1976), *Proc. Natl. Acad. Sci. USA* **73**:3083.

97. Novak, U., Dilworth, S.M., and Griffin, B.E. (1980), *Proc. Natl. Acad. Sci. USA* **77**:3278.

98. Israel, M.A., Simmons, D.T., Mourihan, S.L., Rowe, W.P., and Martin, M.A. (1979), *Proc. Natl. Acad. Sci. USA* **76**:3713.

99. Moore, J.L., Chowdhury, K., Martin, M.A., and Israel, M.A. (1980), *Proc. Natl. Acad. Sci. USA* **77**:1136.

100. Gaudray, P., Rassoulzadegan, M., and Cuzin, F. (1978), *Proc. Natl. Acad. Sci. USA* **75**:4987.

101. Goldman, N., Brown, M., and Khoury, G. (1981), *Cell* **24**: 567.

102. Pöckl, E. and Wintersberger, E. (1980), *J. Virol.* **35**:8.

103. Basilico, C., Gattoni, S., Zouzias, D., and DellaValle, G. (1979), *Cell* **17**:645.

104. Colantuoni, V., Dailey, L., and Basilico, C. (1980), *Proc. Natl. Acad. Sci. USA* **77**:3850.

105. DellaValle, G., Fenton, R.G., and Basilico, C. (1981), *Cell* **23**:347.

106. Botchan, M., Topp, W.C., and Sambrook, J. (1979), *Cold Spring Harbor Symp. Quant. Biol.* **44**:709.

107. Reed, S.I., Ferguson, J., Davis, R.W., and Stark, G.R. (1975), *Proc. Natl. Acad. Sci. USA* **72**:1605.

108. Prives, C., Beck, Y., and Shure, H. (1980), *J. Virol.* **33**:689.

108a. Klein, G. (1981), *Advances in Viral Oncology*, vol. 2, Raven Press, New York.

109. Tjian, R. (1978), *Cell* **13**:165.

110. Tjian, R. and Robbins, A. (1979), *Proc. Natl. Acad. Sci. USA* **76**:610.

111. Gaudray, P., Clertant, P., and Cuzin, F. (1980), *Eur. J. Biochem.* **109**:553.

112. Griffin, J.D., Spangler, G., and Livingston, D.M. (1979), *Proc. Natl. Acad. Sci. USA* **76**:2610.

113. Anderson, J.L., Chang, C., Mora, P.T., and Martin, R.G. (1977), *J. Virol.* **21**:459.

114. Anderson, J.L., Martin, R.G., Chang, C., Mora, P.T., and Livingston, D. M. (1977), *Virology* **76**:420.

114a. Gilden, R.V., Carp, R.I., Taguchi, F., and Defendi, V. (1965), *Proc. Natl. Acad. Sci. USA* **53**: 684

114b. Khandjian, E.W., Loche, M., Darlix, J.-L., Cramer, R., Türler, H., and Weil, R. (1982), *Proc. Natl. Acad. Sci. USA* **79**: 1139.

115. Schaffhausen, B.S. and Benjamin, T.L. (1976), *Proc. Natl. Acad. Sci. USA* **73**:1092.

116. Seif, R. and Martin, R.G. (1979), *J. Virol.* **32**:979.

117. Martin, R.G., Setlow, V.P., Edwards, C.A.F., and Vembu, D. (1979), *Cell* **17**:635.

118. Hassell, J.A., Topp, W.C., Rifkin, D.B., and Moreau, P.E. (1980), *Proc. Natl. Acad. Sci. USA* **77**:3978.

119. Schlegel, R. and Benjamin, T.L. (1978), *Cell* **14**:587.

120. Graessmann, A., Graessmann, M., Tjian, R., and Topp, W.C. (1980), *J. Virol.* **33**:1182.

121. Smith, A.E., Smith, R., Griffin, B., and Fried, M. (1979), *Cell* **18**:915.

122. Eckhart, W., Hutchinson, M.A., and Hunter, T. (1979), *Cell* **18**:925.

123. Schaffhausen, B.S. and Benjamin, T.L. (1979), *Cell* **18**:935.

123a. Walter, G., Hutchinson, M.A., Hunter, T., and Eckhart, W. (1982), *Proc. Natl. Acad. Sci. USA* **79**: 4025

124. Kasamatsu, H. and Nehorayan, A. (1979), *J. Virol* **32**:648.

125. Mullarkey, M.F., Hruska, J.F., and Takemoto, K.K. (1974), *J. Virol.* **13**:1014.

126. Roblin, R., Härle, E., and Dulbecco, R. (1971), *Virology* **48**:49.

127. Sheehafer, J.G. and Weil, R. (1974), *Virology* **58**:75.

128. Shah, K.V., Ozer, H.L., Ghazey, H.N., and Kelly, T.J., Jr. (1977), *J. Virol.* **21**:179.

128a. Jay, G., Nomura, S., Anderson, C.W., and Khoury, G. (1981), *Nature* **291**: 346

128b. Jackson, V. and Chalkley, R. (1981), *Proc. Natl. Acad. Sci. USA* **78**: 6081.

128c. Barkan, A. and Mertz, J. (1981), *J. Virol.* **37**: 730.

129. Berget, S.M., Moore, C., and Sharp, P.A. (1977), *Proc. Natl. Acad. Sci. USA* **74**:3171.

130. Aloni, Y., Dhar, R., Laub, O., Horowitz, M., and Khoury, G. (1977), *Proc. Natl. Acad. Sci. USA* **74**:3686.

131. Mathis, D.J. and Chambon, P. (1981), *Nature* **290**:310.

132. Gruss, P., Dhar, R., and Khoury, G. (1981), *Proc. Natl. Acad. Sci. USA,* **78**: 943.

133. Sekikawa, K. and Levine, A.J. (1981), *Proc. Natl. Acad. Sci. USA* **78**:1100.

134. Berk, A.J. and Sharp, P.A. (1978), *Proc. Natl. Acad. Sci. USA* **75**:1274.

135. Manaker, R.A., Khoury, G., and Lai, C.-J. (1979), *Virology* **97**:112.

136. Seif, I., Khoury, G. and Dhar, R. (1979), *Nucleic Acids Res.* **6**:3387.

137. Deppert, W., Walter, G., and Linke, H. (1977), *J. Virol.* **21**:1170.

138. Chang, C., Martin, R.G., Livingston, D.M., Luborsky, S.W., Hu, C.-P., and Mora, P.T. (1979), *J. Virol.* **29**:69.

139. Israel, M.A., Martin, M.A., Miyamura, T., Takemoto, K.K., Rifkin, D., and Pollack, R. (1980), *J. Virol.* **35**:252.

140. Tevethia, S.S., Katz, M., and Rapp, F. (1965), *Proc. Soc. Exp. Biol. Med.* **119**:896.
141. Levine, A.S., Oxman, M.N., Henry, P.H., Levin, M.J., Diamandopolus, G.T., and Enders, J.F. (1970), *J. Virol.* **6**:199.
142. Lewis, A.M., Jr. and Rowe, W.P. (1971), *J. Virol.* **7**:189.
143. Robb, J.A. (1977), *Proc. Natl. Acad. Sci. USA* **74**:447.
144. Chang, C., Simmons, D.T., Martin, M.A., and Mora, P.T. (1979), *J. Virol.* **31**:463.
145. Kress, M., May, E., Cassingena, R., and May, P. (1979), *J. Virol.* **31**:472.
146. Yang, Y.-C., Hearing, P., and Rundell, K. (1979), *J. Virol.* **32**:147.
147. McCormick, F. and Harlow, E. (1980), *J. Virol.* **34**:213.
148. Simmons, D.T., Martin, M.A., Mora, P.T., and Chang, C. (1980), *J. Virol.* **34**:650.
149. Simmons, D.T. (1980), *J. Virol.* **36**:519.
150. Gurney, E.G., Harrison, R.O., and Fenno, J. (1980), *J. Virol.* **34**:752.
151. Rundell, K., Major, E.O., and Lampert, M. (1981), *J. Virol.* **37**:1090.
152. Spangler, G.J., Griffin, J.D., Rubin, H., and Livingston, D.M., (1980), *J. Virol.* **36**:488.
153. Sabin, A.B. (1965), *Viruses and Cancer, A Lecture.* Shenval Press, London.
154. Cairns, J. (1978), *Cancer: Science and Society.* W.H. Freeman & Co., San Francisco.
155. Chou, J.Y. (1978), *Proc. Natl. Acad. Sci. USA* **75**:1409.
156. Schlegel-Haueter, S.E., Schlegel, W., and Chou, J.Y. (1980), *Proc. Natl. Acad. Sci. USA* **77**:2731.
157. Rouse, P. (1911), *J. Exp. Med.* **13**:397.
158. Earle, W.R., Schilling, E.L., Stark, T.H., Strauss, N.P., Brown, M.F., and Shelton, E. (1943), *J. Natl. Cancer Inst.* **4**:165.
159. Lo, W.H.Y., Gey, G.O., and Shapras, P. (1955), *Bull. Johns Hopkins Hosp.* **97**:248.
160. Hynes, R.O. (Ed.) (1979), *Surfaces of Normal and Malignant Cells.* John Wiley & Sons, Chichester.
161. Epifanova, O.I. and Terskikh, V.V. (1969), *Cell Tissue Kinet.* **2**:75.
162. Stoker, M.G.P. (1973), *Nature* **246**:200.

163. Holley, R.W. and Kiernan, J.A. (1974), *Proc. Natl. Acad. Sci. USA* **71**:2908.

164. Holley, R.W., Armour, R., and Baldwin, J.H. (1978), *Proc. Natl. Acad. Sci. USA* **75**:1864.

165. Abercrombie, M. (1970), *In Vitro* **6**:128.

166. Bell, P.B. (1977), *J. Cell. Biol.* **74**:963.

167. Macpherson, I. (1969), in *Fundamental Techniques in Virology* (K. Habel and N.P. Salzman, Eds.), p. 214, Academic Press, New York.

168. Rous, P. (1935), in *The Harvey Lectures* (1936), William & Wilkins, Co., Baltimore.

169. Sambrook, J., Westphal, H., Srinivasan, P.R., and Dulbecco, R. (1968), *Proc. Natl. Acad. Sci. USA* **60**:1288.

170. Weinberg, R.A. (1980), *Ann. Rev. Biochem.* **49**:197.

171. Botchan, M., Topp, W., and Sambrook, J. (1976), *Cell* **9**:269.

172. Wold, W.S.M., Green, M., Mackey, J.K., Martin, J.D., Padgett, B.L., and Walker, D.L. (1980), *J. Virol.* **33**:1225.

173. Howley, P.M. and Martin, M.A. (1977), *J. Virol.* **23**:205.

174. Chenciner, N., Grossi, M.P., Menequzzi, G., Corallini, A., Manservigi, R., Barbanti-Brodano, G., and Milanesi, G. (1980), *Virology* **103**:138.

175. Chenciner, N., Meneguzzi, G., Corallini, A., Grossi, M.P., Grassi, P., Barbanti-Brodano, G., and Milanesi, G. (1980), *Proc. Natl. Acad. Sci. USA* **77**:975.

176. Yogo, Y., Furuno, A., Watanabe, S., and Yoshiike, K. (1980), *Virology* **103**:241.

177. Schegget, J.T., Voves, J., VanStrien, A., and VanderNoordaa, J. (1980), *J. Virol.* **35**:331.

178. Fresen, K.O., Cho, M.-S., and zurHausen, H. (1980), in *Viruses in Naturally Occurring Cancers* (M. Essex, G. Todaro, and H. zurHausen, Eds.), pp. 35–44, Cold Spring Harbor Laboratory, Cold Spring Harbor, New York.

179. Stevens, J. (1978), in *Persistent Viruses* (J. Stevens, G.J. Todaro, and C.F. Fox, Eds.), p. 701, Academic Press, New York.

180. Shaw, J.E., Levinger, L.F., and Carter, C.W., Jr. (1979), *J. Virol.* **29**:657.

181. Dhruva, B.R., Shenk, T., and Subramanian, K.N. (1980), *Proc. Natl. Acad. Sci. USA* **77**:4514.

182. Wiche, G., Furtner, R., Steinhaus, N., and Cole, R.D. (1979), *J. Virol.* **32**:47.

183. Perbal, B. (1980), *J. Virol.* **35**:420.

184. Seif, R. (1980), *J. Virol.* **35**:479.

185. Soprano, K.J., Rossini, M., Croce, C., and Baserga, R. (1980), *Virology* **102**:317.

186. Friedmann, T., Doolittle, R.F., and Walter, G. (1978), *Nature* **274**:291.

187. Lewis, W.H. (1935), *Science* **81**:545.

188. Wolman, S.R., Hirschhorn, K., and Todaro, G.J., (1964), *Cytogenetics* **3**:45.

189. Theile, M. and Strauss, M. (1977), *Mutation Res.* **45**:111.

190. Theile, M., Scherneck, S., and Geissler, E. (1980), *Arch. Virol.* **65**:293.

191. Hirai, K., Campbell, G., and Defendi, V. (1974), in *Control of Proliferation in Animal Cells* (B. Clarkson and R. Baserga, Eds.), p. 151, Cold Spring Harbor Laboratory, Cold Spring Harbor, New York.

192. Vogel, A. and Pollack, R. (1974), *J. Virol.* **14**:1404.

193. Fenner, F. (1968), *The Biology of Animal Viruses. Volume II; The Pathogenesis and Ecology of Viral Infections.* Academic Press, New York.

194. Wagner, R.R. (1961), *Virology* **13**:323.

195. Sanders, M. and Schaeffer, M. (Eds.) (1971), *Viruses Affecting Man and Animals.* Warren H. Green, St. Louis.

196. Younger, J.S. and Preble, O.T. (1980), in *Comprehensive Virology—Virus-Host Interactions: Viral Invasion, Persistence and Disease* (H. Fraenkle-Conrat and R.R. Wagner, Eds.), Vol. 16, Plenum Press, New York.

197. Sigurdsson, B., Pálsson, P.A., and Grímsson, H. (1957), *J. Neuropathol. Exp. Neurol.* **16**:389.

198. Holland, J.J., Villarreal, L.P., and Etchinson, J.R. (1974), in *Mechanisms of Virus Disease* (W.S. Robinson and C.F. Fox, Eds.), Vol. 1, p. 131. W.A. Benjamin, Menlo Park.

199. Robb, J.A. (1978), in *Human Diseases Caused by Viruses* (H. Rothschild, F. Allison, Jr., and C. Howe, Eds.), p. 45, Oxford University Press, New York.

200. Johnson, R.T., Narayan, O., and Weiner, L.P. (1974), in *Mecha-*

nisms of Virus Disease (W.S. Robinson and C.F. Fox, Eds.), Vol. 1, p. 187, W.A. Benjamin, Menlo Park.

201. Huang, A.S. and Baltimore, D. (1970), *Nature* **226**:325.

202. Robinson, W.S. (1974), in *Mechanisms of Virus Disease* (W.S. Robinson and C.F. Fox, Eds.), p. 25, W.A. Benjamin, Menlo Park.

203. Norkin, L.C. (1976), *Infect. Immun.* **14**:783.

204. Norkin, L.C. (1979), *Virology* **95**:598.

204a. Grinnell, B.W., Martin, J.D., Padgett, B.L., and Walker, D.L. (1982) *J. Virol.* **43**: 1143.

205. Porter, D.D., Larson, A.E., Cox, N.A., Porter, H.G., and Suffin, S.C. (1977), *Intervirology* **8**:129.

206. Granoff, A. and Naegele, R.F. (1974), in *Mechanisms of Virus Disease* (W.S. Robinson and C.F. Fox, Eds.), Vol. 1, p. 493, W.A. Benjamin, Menlo Park.

207. Wechsler, S.L., Rustigan, R., Stallcup, K.C., Byers, K.B., Winston, S.H., and Fields, B.N. (1979), *J. Virol.* **31**:677.

208. Schmaljohn, C.S. and Blair, C.D. (1979), *J. Virol.* **31**:816.

209. Doller, E., Aucker, J., and Weissbach, A. (1979), *J. Virol.* **29**:43.

210. Fujinami, R.S. and Oldstone, M.B.A. (1979), *Nature* **279**:529.

211. Mocarski, E.S. and Stinski, M.F. (1979), *J. Virol.* **31**:761.

212. Pauw, W. and Choufoer, J. (1978), *Arch. Virol.* **57**:35.

213. Lecatsas, G., Prozesky, O.W., VanWyk, J., and Els, H.J. (1973), *Nature* **241**:343.

214. Jung, M., Krech, U., Price, P.C., and Pyndiaii, M.N. (1975), *Arch. Virol.* **47**:39.

215. Soriano, F., Shelburne, C.E., and Gökcen, M. (1974), *Nature* **249**:421.

216. Weiss, A.F., Portman, R., Fischer, H., Simon, J., and Zang, K.D. (1975), *Proc. Natl. Acad. Sci. USA* **72**:609.

217. Coleman, D.V., Daniel, R.A., Gardner, S.D., Field, A.M., and Gibson, P.E. (1977), *Lancet* **2**:709.

218. Taguchi, F., Nagaki, D., Saito, M., Haruyama, C., Iwasaki, K., and Suzuki, T. (1975), *Jpn. J. Microbiol.* **19**:395.

219. Takemoto, K.K. and Mullarkey, M.F. (1973), *J. Virol.* **12**:625.

220. Dougherty, R.M. and diStephano, H.S. (1974), *Proc. Soc. Exp. Biol. Med.* **146**:481.

221. Padgett, B.L. and Walker, D.L. (1976), *Prog. Med. Virol.* **22**:1.

222. Gardner, S.D. (1977), *Recent Adv. Clin. Virol.* **1**:93.

223. Gardner, S.D. (1973), *Br. Med. J.* **1**:77.

224. Padgett, B.L. and Walker, D.L. (1973), *J. Infect. Dis.* **127**:467.

225. Padgett, B.L., Rogers, C.M., and Walker, D.L. (1977), *Infect. Immun.* **15**:656.

226. ZuRhein, G.M., Padgett, B.L., Walker, D.L., Chun, R.W.M., Horowitz, S.D., and Hong, R. (1978), *N. Engl. J. Med.* **299**:256.

227. Holmberg, C.A., Gribble, D.H., Takemoto, K.K., Howley, P.M., Espana, C., and Osburn, B.I. (1977), *J. Infect. Dis.* **136**:593.

228. Costa, J., Yee, C., and Rabson, A.S. (1977), *Lancet* **ii**:709.

229. Corallini, A., Barbanti-Brodano, G., Portolani, M., Balboni, P.G., Grossi, M.P., Possati, L., Honorati, C., La Placa, M., Mazzoni, A., Capato, A., Veronesi, W., Orefice, S., and Cardinali, G. (1976), *Infect. Immun.* **13**:1684.

230. Pater, M.M., Pater, A., and diMayorca, G. (1979), *J. Virol.* **32**:220.

231. Pater, A., Pater, M.M., and diMayorca, G. (1980), *J. Virol.* **36**:480.

232. Shein, H.M. and Enders, J.F. (1962), *Proc. Soc. Exp. Biol. (N.Y.)* **109**:495.

233. Koprowski, H., Ponten, J.A., Jensen, F., Ravdin, R.G., Moorhead, P., and Saksela, E. (1962), *J. Cell. Comp. Physiol.* **59**:281.

234. Girardi, A.J., Weinstein, D., and Moorhead, P.S. (1965), *Ann. Med. Exp. Biol. Fenn.* **44**:242.

235. Fedoroff, S. and Hertz, L. (Eds.) (1977), *Cell, Tissue, and Organ Cultures in Neurobiology,* Academic Press, New York.

236. Deppert, W. and Walter, G. (1976), *Proc. Natl. Acad. Sci. USA* **73**:2505.

237. Walter, G., Scheidtmann, K.-H., Carbone, A., Laundano, A.P., and Doolittle, R.F. (1980), *Proc. Natl. Acad. Sci. USA* **77**:5197.

238. Friedman, R.M. (1977), *Bacteriol. Rev.* **41**:543.

239. Kingsman, S.M., Smith, M.D., and Samuel, C.E. (1980), *Proc. Natl. Acad. Sci. USA* **77**:2419.

240. Kingsman, S.M. and Samuel, C.E. (1980), *Virology* **101**:458.

241. Mozes, L.W. and Defendi, V. (1979), *Virology* **93**:558.

242. Mondal, S., Brankow, D.W., and Heidelberger, C. (1976), *Cancer Res.* **36**:2254.

243. Kinsella, A.R. and Radman, R. (1978), *Proc. Natl. Acad. Sci. USA* **75**:6149.

244. Klein, G. (1979), *Proc. Natl. Acad. Sci. USA* **76**:2442.

245. Seif, R. (1980), *J. Virol.* **36**:421.

246. Castro, J.E. (Ed.) (1978), *Immunological Aspects of Cancer,* University Park Press, Baltimore.

247. Turk, J. (Ed.), (1977), *Current Topics in Immunology: Immunodeficiency,* Williams & Wilkins Co., Baltimore.

HERPESVIRUSES

I. INTRODUCTION

The herpesviruses are usually defined as particles of about 100 nm in diameter that bud through the inner membrane of the infected cell, have DNA as their genetic material being packaged into capsid containing 162 capsomers, and are enclosed in an envelope. It is clear that such a broad definition could not lead to the constitution of an homogeneous taxonomic group, but rather to the collection of a wide variety of different viruses that may share very few, if any, genetic homology. Thus, more than 70 distinct herpesviruses have been described in almost all species of eucaryotic cells, including those of fish, amphibians, reptiles, birds, and mammals. The great diversity of the group is illustrated by the virus names, which refer to their natural host (e.g., squirrel monkey herpesvirus, tree shrew herpesvirus, channel catfish herpesvirus, canine herpesvirus, *Herpesvirus hominis*), or to the name of their discoverer (e.g., Epstein-Barr virus, Lucké virus), or to the disease that they cause (e.g., varicella zoster, cytomegalovirus, equine abortion virus, feline infectious laryngotracheitis, bovine mammilitis).

Our rapidly expanding knowledge in molecular biochemistry and cellular biology will allow us to use more restrictive criteria than those originally provided by ultrastructural studies to define subclasses in the herpesvirus group. One of the prominent points of interest in the study of the biology of the herpesvirus is that several members of the group have been associated with animal and human cancers—for example, Epstein-Barr virus; *H. hominis, H. saimiri*, the squirrel monkey virus; the herpes simplex virus; Lucké virus; and Marek's disease virus. The major part of our current knowledge in the biology of herpesviruses emerges from studies of simplex virus type 1 and 2, Epstein-Barr virus, and, more recently, cytomegalovirus, pseudorabies virus, and *H. saimiri*. The following sections deal mainly with herpes simplex virus biology, which is being clarified at a remarkable rate.

II. PHYSICAL AND GENETIC PROPERTIES

A. ULTRASTRUCTURE OF THE HERPES VIRION

The herpes virion is composed of a core surrounded by a capsid, tegument, and a membrane (1,2). The structure of herpes simplex type 1 (HSV_1) virus is schematically drawn in Fig. 1. The core of the virus (25–30 nm ϕ), which resembles a toroid, contains DNA (3,4,5,6,7,8) wound around a cylindrical structure (8). The capsid has an outer dimension of about 100 nm and has been subdivided in three components (9): an inner capsid, a middle capsid, and an outer capsid, which are 10 nm, 15nm, and 12 to 15 nm thick, respectively. The outer capsid

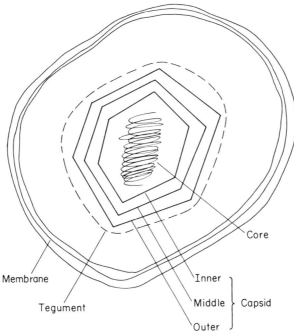

FIGURE 1.   Diagrammatic representation of the herpes simplex virus virion structure.

has been described as an icosahedral structure consisting of 162 capsomers arranged in a 5:3:2 symmetry and consisting of elongated prisms of 92 to 125 nm diameter with a 40 nm axial hole (10). The tegument is located between the outer capsid and the membrane. This structure, which has been shown to vary greatly in size among the different herpesviruses (11,12,13,14,15), appears slightly denser than the adjacent capsid or inner lamella of the envelope and has been described as showing a fibrous appearance when negatively stained (2). The membrane of the virion is the most external component of the particle (15,16,17,18,19,5) and has a trilaminar structure with spikes of about 80 to 100 Å long, being apart about 50 Å from the surface of the envelope (10,20,21,22). Involvement of the virus membrane in the infectious process is suggested by the observations that naked particles are not infectious (23,24,25), enveloped particles being twice as likely to be adsorbed as naked particles (16,26). It is likely, therefore, that the presence of a membrane is required for efficient cell penetration, which has been described as occurring either by pinocytosis (16,26,27) or by fusion of the plasma membrane (28). The virion membrane probably originates from the cell membrane during the release of viral particles which is thought to proceed by specific channels (29,30,31) although reverse pinocytosis has also been described (32).

B.   BIOCHEMICAL COMPOSITION OF THE HERPES VIRIONS

The components of herpes virions are proteins, lipids, polyamines, and DNA.

1.   Virion Polypeptides

The identification of the structural proteins constituting the intact viral particles requires an extensive purification of the virions in order to avoid a possible contamination by cellular proteins, which may be associated with the membrane vesicles that are generated during virus purification. The use of several different methods to purify herpes simplex virions (33–40) and the availability of techniques allowing high-resolution electrophoresis of proteins on polyacrylamide gels (41,43) have allowed to establish that HSV particles contain at least 33 different polypeptide species whose molecular weight ranges from 11,000 to 275,000 daltons (38,42,44–47). The composition of several other herpesviruses have been studied in the same fashion and has revealed the existence of several similarities. It has been, therefore, shown that most of the herpesviruses tested thus far are composed of about 20 to 34 polypeptides (48–50) and that about 4 to 10 of them are glycosylated. Comparative studies performed with herpesviruses from different origins (51–60) have allowed recognition of the presence of a major high-molecular-weight (~150,000 daltons) capsid polypeptide in all the virions tested. The reader interested in these studies should consult the reviews by Honess and Watson (61) and Roizman and Spear (62).

2. Lipids and Polyamines

The presence of lipids and polyamines (spermine and spermidine) has already been described in preceding reviews (2,62). We shall mention here only that spermine seems to be associated with the nucleocapsid whereas spermidine is localized in the tegument or envelope, and that the presence of lipids is necessary to insure an adequate infectivity of the virion (63).

3. Herpesvirus DNA

Herpesvirus DNA is a large linear double-stranded molecule characterized by a high G + C content and the presence of

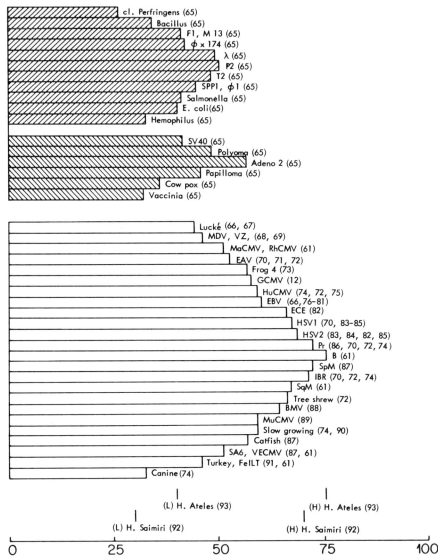

FIGURE 2. Base composition of herpesvirus. The G and C content (expressed as %) of herpesvirus DNAs (open rectangles) is compared with that of some procaryotes *(Hemophilus, Escherichia coli, Salmonella, Clostridium perfringens, Bacillus)*, bacteriophages (F1, M13, ϕ×174, v, P_2, T_2, SPP_2, ϕ_1), and some animal viruses (SV40, polyomavirus, adenovirus, papillomavirus, cowpox, vaccinia). References are cited in parentheses after the name of the organism or virus.

stretches of repetitive sequences, bracketing one or two unique sequences of varying size.

α. **Base Composition.** The mean nucleotide composition of herpesvirus DNAs has long been considered a valuable taxonomic criterion (64,65), because many of the herpes DNAs exhibit a high content of guanine + cytosine (mean 60%) as compared to the values obtained for other eucaryotic viruses or procaryotes and their viruses.

Attempts have been made to distinguish several subgroups among the herpesviruses on the basis of their G + C content. However, no clearcut relationship has been established between the different members of such classes. Therefore, it has been shown that viruses having very different G + C content may share common properties, whereas apparently unrelated viruses may have very close G + C contents. The variability observed might be the result of the deletion or the duplication of small DNA regions whose G + C content is particularly high. It has been shown, for example, the *H. ateles* and *H. saimiri* exhibit an unusual heterogeneity in base composition (92,93), such that repetitive regions of their DNA molecule contain as much as 75% G + C, whereas the unique region contains only 35% G + C (67,69,70).

b. **Structure of Herpesvirus DNA**

☐ i. PURIFICATION OF HERPESVIRUS DNA. *The determination of herpesvirus DNA structure relies on the isolation of intact molecules from the virions. Several procedures have been devised to purify intact viral DNA from herpesvirus-infected cells (43,94–96). They generally include a mechanical disruption of the infected cells, purification of the virions through a sucrose gradient, and isolation of herpes DNA on the basis of its high buoyant density as compared with that of cellular DNA.*

Virions can also be obtained after centrifugation of the culture medium in which particles are released from in-

fected cells. Purification of the DNA is generally achieved by phenol extraction. Alternatively, the cells may be lysed with sodium dodecylsulfate (SDS) to a final concentration of 0.6%. After digestion of the proteins by pronase, the extract is centifuged in a sodium iodide density gradient in the presence of ethidium bromide (30 µg/ml), allowing the direct visualization of DNA by ultraviolet illumination. This method allows the isolation of large quantities of herpes DNA (97). Full-length viral DNA molecules can also be obtained from herpes-simplex-infected cells in the presence of detergent at high tonic strength (0.2M NaCl) (98), whereas equine abortion virus (EAV) DNA has been separated from cellular DNA by the employment of methylated-albumin Kieselguhr (MAK) column chromatography (Fig. 3) (99). □

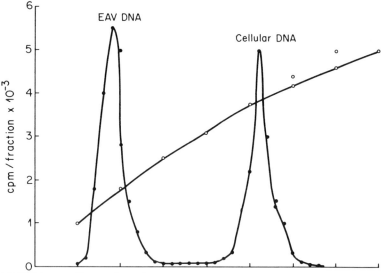

FIGURE 3.   Separation of cellular and EAV DNAs by MAK chromatography. Thirty µg of ^{14}C-labeled cellular DNA and 58 µg of ^{14}C-labeled EAV DNA have been chromatographed with a 0.4 to 0.8 *M* NaCl gradient (open circles). The recovery of DNA is 93.7% for EAV and 95% for cell DNA (99).

ii. MOLECULAR WEIGHT OF HERPESVIRUS DNA. Herpes DNA is a linear double-stranded molecule of about 90 to 100 10^6 daltons in molecular weight [pseudorabies virus: 90×10^6 daltons (100), Epstein-Barr virus: 105×10^6 daltons (81), Marek's disease virus: 103×10^6 daltons (68), equine abortion virus: 94×10^6 daltons, (71), and tree shrew *Herpesvirus:* 100×10^6 daltons (101)]. Two significantly different values have been reported for murine cytomegalovirus: 132×10^6 daltons (89) and channel catfish virus: 84×10^6 daltons (102). The molecular weight of herpes simplex virus has been estimated to be 95 to 100×10^6 daltons from measurement of the contour length of the molecule (103–105), sedimentation velocity in sucrose gradients (84), sequence complexity (106), and summation of the size of restriction endonuclease fragments (107).

iii. STRUCTURE OF HERPES SIMPLEX VIRUS (HSV) DNA. Schematically, the molecule of HSV DNA consists of two covalently linked components, comprising respectively, 82% and 18% of the viral DNA. Each of the two components, called long (L) and short (S), consists of unique sequences (U_L and U_S) bracketed by inverted repeats (IR) comprising about 5 to 6% of the molecule length (Fig. 4).

Under self-annealing conditions, the alkali-denatured molecules give rise to the formation of two single-stranded loops of unequal size bridged by double-stranded DNA (110) (Fig. 5), and upon digestion with lambda exonuclease, followed by annealing at high temperature (65°C), the molecules of HSV DNA can circularize, due to the unmasking of cohesive ends (104) (Fig. 6). A minimum of 400 base pairs has been reported to be necessary at both ends to allow circularization (111), whereas a

FIGURE 4. Diagrammatic representation of herpes simplex virus DNA.

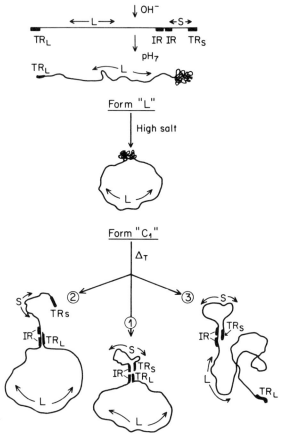

FIGURE 5.   Possible scheme for the generation of L, C_1, and C_2 forms of intact HSV DNA strands. From Sheldrick and Berthelot (110).

repeat unit of about 280 base pairs has been mapped at both termini and L-S junction (112).

The reiterated sequences bracketing the L fragment have been called ab and b'a'; those bracketing the S region (which have been called a'c' and ca) represent 4.3% of the HSV DNA molecule. The different sensitivity of these redundant sequences to restriction nucleases suggest that ab, ac, and a'b', a'c' sequences

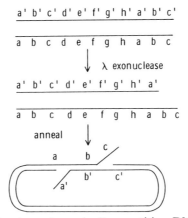

FIGURE 6. Digestion of terminally repetitive DNA molecule with exonuclease, exposing complementary single strands which reassociate under conditions favoring DNA renaturation.

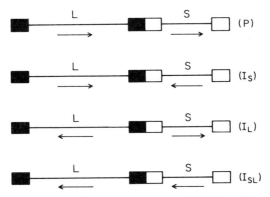

FIGURE 7. Possible different arrangements of the long (L) and short (S) unique sequences in HSV DNA. Inversion of the long and short sequences give rise to I_L, I_S, and I_{SL} arrangements. The prototype arrangement is designed P.

are different (108,109,113–116). A similar conclusion was obtained from reassociation studies (105), whereas recent studies with mutants (see below) lead to the conclusion of obligatory identity for the terminal portions of the reiterated sequences in

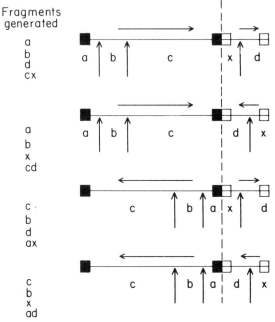

Fragments
generated

FIGURE 8.   DNA fragments generated after digestion of the four arrangements of HSV DNA with endonucleases that cleave outside of reiterated regions (open and closed boxes).

the S component and of the "a" sequence in both L and S components.

The presence of complementary inverted sequences at the ends and within the DNA molecule of HSV allowed Sheldrick and Berthelot (110) to predict the existence of four isomers which would differ by the relative orientations of the L and S components (Fig. 7) arising after recombinations.

The existence of four distinct DNA populations in purified virions has been confirmed by the establishment of restriction endonuclease cleavage maps (108,115,118) and studies of partial denaturation profiles (108,117). These four arrangements have been called P (prototype), I_L (inversion of L), I_S (inversion of S), and I_{SL} (inversion of both S and L) (108, 118). The ratio-

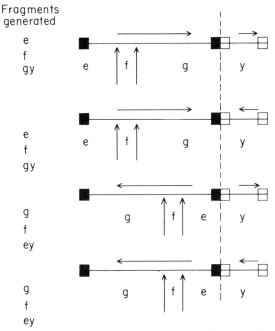

FIGURE 9. DNA fragments generated after digestion of the four arrangements of HSV DNA with endonucleases that cleave only one set of unique region.

nale used along the construction of the physical maps is detailed below.

☐ *Let us consider a fragment F formed by digestion with the restriction enzyme E_1 in a molar ratio with respect to the concentration of the DNA molecule.*

Let us call $F_{1/1}$ and $F_{1/2}$ the fragments generated by the digestion of F_1 with a second restriction enzyme, E_2.

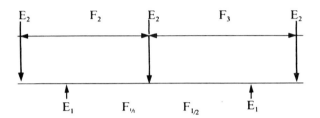

The neighboring restriction sites of enzyme E_2 delineate fragments F_2 and F_3. These fragments will be cut by enzyme E_1 to give $F_{2/1}$, $F_{2/2}$, and $F_{3/1}$, $F_{3/2}$.

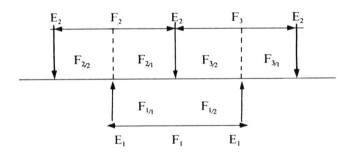

Fragments $F_{2/1}$ and $F_{1/1}$ will have the same molecular weight and therefore will allow to locate F_2 contiguously to F_1. The same rationale applied to $F_{3/2}$ and $F_{1/2}$ will lead to the ordering of $F_2/F_1/F_3$.

Application of this logic to all the fragments generated by double digestions has led to the completion of the HSV maps. An example is illustrated in Table I.

The relative position of the restriction fragments in the map can be deduced directly from Table 1. For example, when fragment D generated by the bacterial restriction en-

TABLE I. Physical Map of HSV DNA

EcoR1 Fragments	Molecular Weight (megadaltons)	Molar Ratio	Molecular Weight of Hind III Redigestion Products
A	13.7	1.0	4.1–6.0–3.6
B_1, B_2	13.5	0.5	13.5
C_1, C_2	11.5	0.5	11.5
D	10.4	1.0	0.8–7.9–1.7
E	10.2	0.5	10.2
F	10.2	1.0	10.2
G	10.2	1.0	3.5–6.7
H	9.6	1.0	5.4–3.0–1.20
I	8.5	1.0	7.3–1.25
J	8.2	0.5	8.2
L	3.4	1.0	3.4
K_1, K_2	3.3	1.0	3.3
M	2.5	1.0	2.5
N	1.4	1.0	1.4
O	0.9	1.0	0.9

Hind III Fragments	Molecular Weight (megadaltons)	Molar Ratio	Molecular Weight of EcoR1 Redigestion Products
B	26.2	0.25	7.3–13.5–5.4
A	25.5	1.0	3.5–1.4–10.2–2.5–0.9–3.4–3.6
C	22.0	0.25	7.3–13.5–1.20
D	17.5	0.5	10.2–7.3
E	16.9	0.25	11.5–5.4
F	12.7	0.25	11.5–1.20
G	8.2	0.5	5.4–3.3
H	7.9	1.0	8.2
I	7.5	1.0	7.9
J	6.0	1.0	6.7–0.8
K	5.4	1.0	6.0
L	4.5	0.5	1.25–4.1
M	3.0	1.0	1.20–3.3
N	1.7	1.0	3.0
O			1.7

*donuclease EcoR1 is redigested with Hind III, another re-
striction endonuclease, three fragments are obtained.*

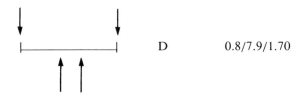

D   0.8/7.9/1.70

*One of these fragments is necessarily found among the
Hind III products of DNA digestion. This is the case with
7.9 and 1.7. Hind III digest gives rise only to a 0.8 fragment when redigested with EcoR1. This fragment must,
therefore, overlap EcoR1 D.*

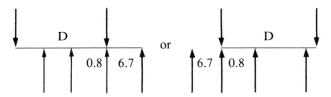

Only one EcoR1 fragment (G) can overlap Hind III 6.7,

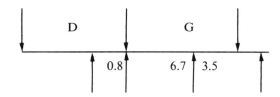

*and the only Hind III fragment that can overlap fragment
3.5 is A.*   □

One of the consequences of the unique structure of HSV
DNA is that endonucleases that do not cleave within the reiter-
ated regions will generate three classes of fragments (Fig. 8).
The first class corresponds, in concentration, to the molarity of

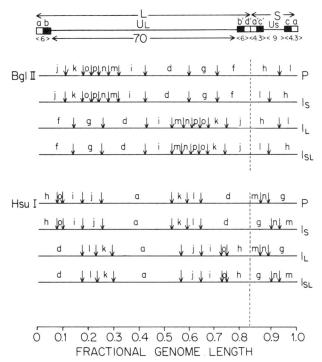

FIGURE 10. Different fragments generated after HSV₁ DNA digest with Bgl II and HsuI restriction endonucleases. The fragments shown are obtained with the four different orientations of the long and short unique region of HSV DNA.

the intact DNA (four fragments b, four DNA arrangements). These molar fragments map between the first and the last cleavage site within the unique L region (and within the unique S region). The second class comprises the terminal fragments of the L and S regions (a,c,d,x). Each of them is represented twice in the four arrangements and therefore, are, called half molar. The third class comprises the fragments resulting from the junction of terminal fragments of the L and S regions as a consequence of the inversion of UL and US. These fragments are

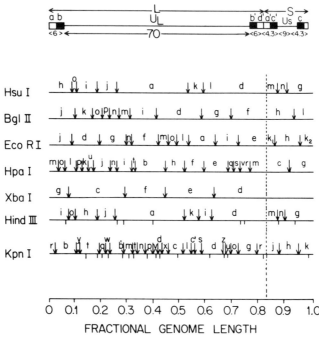

FIGURE 11.   Maps of some restriction sites on HSV₁ DNA.

represented only once for the four arrangements and are called quarter molar (cx,cd,az,ad).

Enzymes that cleave within only one set of unique regions do not produce quarter molar fragments (Fig. 9).

Several different restriction endonucleases have been used recently to establish detailed maps (see Figs. 10–13). The designation of the fragments follows a convention proposed by Jones et al. (119) in which DNA fragments are designated by letters of the alphabet in order of decreasing size. The junction fragments are designated either by the two letters identifying the individual fragments from which they are composed or by an appropriate letter. The joint region is that point where the inverted repetitions of the termini of HSV DNA molecule meet.

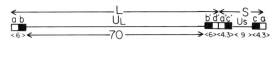

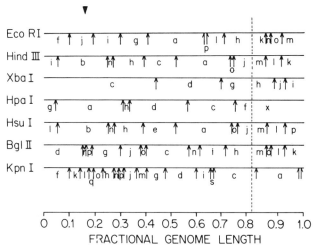

FIGURE 12. Maps of some restriction sites on HSV₂ DNA.

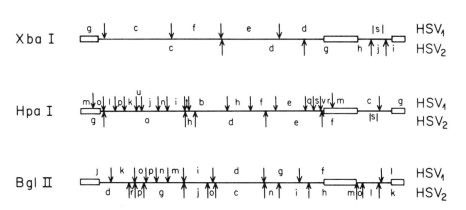

FIGURE 13. Comparative maps for some restriction sites on HSV₁ and HSV₂ DNA.

Arrangement of the origin of double digest fragments from individual single digest fragments may be achieved:

(a)   by further cleavage of isolated fragments;
(b)   by hybridization of ³²P-labeled fragments produced with one restriction enzyme to the unlabeled fragments produced by a different restriction enzyme;
(c)   by direct comparison of the independent single digest cleavage patterns produced by two enzymes with the double digest pattern.

Fine-structure restriction endonuclease cleavage site mapping of the joint region has revealed the presence of insertions of 280 base pairs or multiples of 280 base pairs being probably tandem duplications of sequences spanning the joint and corresponding closely to the inverted terminal repeats in about 50% of the DNA molecules from all plaque-purified virus (112). Restriction endonuclease cleavage sites at the ends of HSV DNA correspond to sites in the inverted repeat regions at the joint, indicating that the sequences of the terminal and internal repeats are probably identical. If the restriction endonuclease cleavage site maps of the ends of HSV DNA molecule are inverted and lined up with the map of the joint region, the ends appear to overlap by 265 base pairs (112). Such an overlapping was also suggested by electron microscopy studies (105,110). The 280 base pair insertions have never been found in the right terminus of HSV, strain KOS, although the same sequences in which such insertions arise in the left terminus and at the joint, i.e., the terminal redundancy, are also present in this region. The presence of tandem duplications of sequences including the terminal redundancy may reflect the presence of a recombinational hot spot, which may also account for the inversion of the L and S fragments (112). Models for the arrangement of repeated sequences in HSV DNA are reported in Fig. 14.

```
 TR                          TR TR'          TR
abcd                        d'c'b'd'b'd'z'y'    yzab   (1)
```

```
 TR                            TR'            TR
abcd                        d'c'b'd'z'y'      yzab   (2)
```

FIGURE 14. Model for arrangement of repeated sequences in HSV DNA. Adapted from Wagner and Summers (112).

iv. STRUCTURE OF EPSTEIN-BARR VIRUS DNA. Epstein-Barr virus (EBV), which is unequivocally the etiologic agent for infecious mononucleosis, has been observed in the childhood Burkitt's lymphoma and appears to be intimately associated with naso-pharyngeal carcinoma. Several isolates of EBV have also been obtained from nonhuman primates.

Primate B lymphocytes are the only cells that have been in-fected with EBV *in vitro* (120,121), as EBV infection is largely nonpermissive. Some lymphoblastoid cell lines infected with EBV *in vitro* or *in vivo* continuously give rise to enough pro-geny to purify the virus and characterize the viral DNA (81,122–127). Three virus isolates have been essentially used

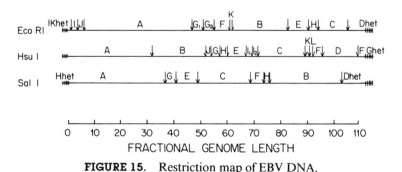

FIGURE 15. Restriction map of EBV DNA.

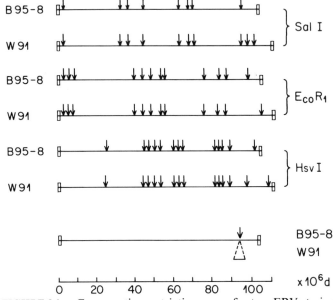

FIGURE 16. Comparative restriction maps for two EBV strains.

thus far to study the structure of EBV DNA. B95-8 was originally obtained from marmoset cells infected *in vitro* with EBV
derived from a continuous culture of lymphoblasts established
from a patient with mononucleosis (128). W-91 was isolated
from a culture obtained from a Burkitt's tumor biopsy (129),
and HR-1 virus was produced by a human lymphoblast cell line
established from a Burkitt's tumor biopsy (130).

The DNA purified from both HR-1 and B95-8 virus cultures
is a linear double-stranded molecule of approximately 100×10^6
daltons, as determined by velocity sedimentation in neutral sucrose gradients (80,81) and by measurement of contour length
(81,131). The buoyant density of this DNA (66,76–78,81,132) is
compatible with a G + C content of 57 to 58% (see Fig. 2). Sedimentation of alkali denature DNA indicates a size less than
50×10^6 daltons for the single strands. Restriction endonuclease
maps have been established (Figs. 15 and 16) and show that

EBV DNA consists of short (10×10^6 daltons) and long (87×10^6 daltons) unique DNA sequences joined by tandem reiterations of a 1.85×10^6 daltons DNA fragment (122,123,126,133) arranged in adjacent tandem array with all copies having the same orientations (134). Part of the internal reiteration (less than 5×10^5 daltons) is also found at two different locations in the long unique region of the DNA molecule and the number of internal random reiterations may vary from 10 to 4 within the different preparations of the same EBV strain (135). Hybridization experiments with purified viral DNA indicate that the DNA of B95-8 strain shares approximately 85 to 92% of the HR-1 strain DNA sequences and that HR-1 virus contains more than 95% of the B95-8 DNA sequences, whereas EBV W-91 DNA contains all of the sequences present in both B95-8 and HR-1 strains (81,125,127). EBV W-91 DNA differs from EBV B95-8 DNA by additional 7×10^6 to 8×10^6 of DNA in the long unique DNA region (135).

At both ends of the molecule are from 1 to 12 repeats of another 3×10^5 dalton sequence (122,136). A very elegant study of these repeats has been reported by Given et al. (136) who showed that: (a) multiple direct repeats are within a terminus, (b) there exists a direct repeat at both ends of the EBV DNA molecule, and (c) the direct repeats at both ends are directly repeated within each end. These observations could be related to the finding that EBV (B95-8) DNA has been found previously in a circular form in cells transformed by this virus (see below and 131,137,138).

The structure of the EBV DNA has been reviewed in great detail by Kieff et al. (139). A diagrammatic representation of EBV DNA is drawn in Fig. 17.

v. STRUCTURE OF OTHER HERPES DNAs. Several herpesviruses from different origins exhibit a genome organization similar to that described above for herpes simplex virus. These include human cytomeglovirus (Hu CMV) and *H. aotus* (night monkey herpesvirus) 1 and 3 (P. Sheldrick, N. Berthelot, M. Laithier,

(a)

(b)

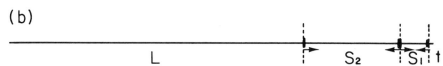

FIGURE 17. Diagrammatic representation of EBV DNA.

and B. Fleckenstein, personal communication), (HVT) (tamarin marmoset herpesvirus) and Marek's disease virus (MDV) (J. Cebrian, C. Kaschka-Dierich, N. Berthelot and P. Sheldrick, personal communication), and bovine mammitis virus (BMV) (140,141).

The DNA of pseudorabies (Pr) virus is a linear double-stranded molecule with a molecular weight of approximately 90 × 10⁶ daltons (100) consisting of a short unique sequence (molecular weight 6 × 10⁶ daltons) bracketed by inverted repeats (molecular weight 9.9 × 10⁶ daltons). The terminal end of the inverted repeat is not found internally, and a long unique sequence (molecular weight 65 × 10⁶ daltons) constitutes the remainder of the molecule (142,143). This particular organization is also shared by equine abortion virus (EAV) DNA, where the long and short regions are 65 × 10⁶ daltons and 6 × 10⁶ daltons long, respectively (P. Sheldrick and N. Berthelot, personal communication).

Another variation in the structural organization of herpesvirus genome is exemplified by the situation encountered with *H. saimiri* and *H. ateles*, both of which are highly oncogenic in primates (144–151). The arrangement of DNA sequences of those two herpesviruses reveals a very tight homology. In both cases, two types of DNA molecules are found: the M genome and the H genomes. The M genome consists of 72 to 74% unique light sequences (L DNA, 36 to 38% of G + C) and 26 to 28% of highly repetitive heavy sequences (H DNA with 71 to 75% G + C). The L DNA sequences (about 70×10^6 daltons) is bracketed, in the M genome, by stretches of H DNA of variable length. On the other hand, the H genome consists of long molecules of highly repetitive DNA sequences that are identical to the terminal H sequences of the M genome (93,152). The repetitive region contains approximately 35 copies of a small tandemly reiterated sequence (1300 base pairs). These repeat units are distributed randomly between the ends of the DNA molecules and, therefore, create a partially circularly permutated population (152,153). Cottontail rabbit herpesvirus (CTHV) has been compared to a "super-saimiri." Its genome appears to consist of two regions of unique sequences (U45 and U60) enclosed by inverted repeats of variable length. It is possible that these repeats consist of random repetition of an 800 base sequence (J. Cebrian, N. Berthelot, M. Laithier, and P. Sheldrick. Abstracts of International Workshop on Herpes Viruses, July 27–31, 1981, Bologna, Italy).

Finally, the simplest DNA sequence arrangement described thus far refers to that of channel catfish virus (CCV), a herpes virus causing a lethal disease in populations of channel catfish fry (154). Velocity sedimentation, electron microscopy, and summation of restriction enzyme fragment's molecule weights reveal that the genome is a linear DNA molecule of 84×10^6 daltons (102). The distribution of restriction endonuclease cleavage sites in the DNA molecule shows that a set of sequences of about 10×10^6 daltons at one end are repeated with direct polarity at the other end. Although local internal reiteration is present, as also reported for EBV (122,126), the maps obtained suggest that the overall genomic sequence order is nonpermuted (102).

It appears, therefore, that five groups of DNA sequence ar-

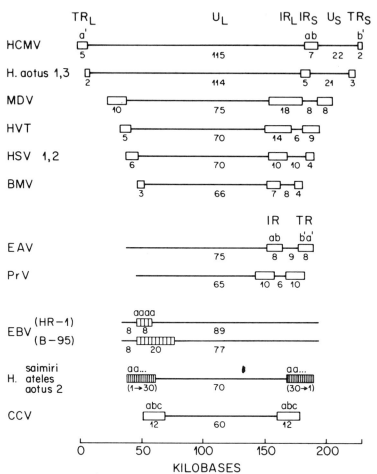

FIGURE 18. Comparative representation of genome organizations of various herpesvirus DNAs.

rangements can be found among herpesviruses. A comparative representation of the different genome organization in herpesviruses is presented in Fig. 18. An important point of the figure is that MDV genome is HSV-like, rather than, as some people had expected, EBV-like or *H. saimiri*-like (S. Sheldrick, personal communication).

The differences in structure and sequence organization observed among the herpes DNA molecules may result either from the wide heterogeneity of the group due to the imprecise taxonomic parameters used to date, or the herpesviruses may be related to each other and have a common ancestor. More additional information of the structure of virus genomes is needed before a decision can be made as to whether "these differences can provide basis for subgrouping or whether they are merely samples from a continuum of eccentric arrangements of palindromic and unique base sequences" (61).

Interestingly, the fragmentation of DNAs upon denaturation with alkali appears to be a common feature for several different herpes viruses (68,81,84,89,155–158). Evidence for the existence of gaps within the molecule of herpes DNA comes from studies performed with lambda exonuclease, which gaps to act at internal sites (111). Hyman et al. (159) have reported *in vitro* repair of these interruptions. It is not yet clear whether the nicks are randomly distributed (143,157,159) or occur at specific sites (156) in the herpesvirus DNA molecule.

III. MULTIPLICATION OF HERPESVIRUSES

The general characteristics of the reproduction cycle of herpesviruses have been described in great detail in other reviews (2, 160–162). We will focus in this chapter only on some aspects of herpesvirus multiplication in mammilian cells.

A. INHIBITION OF HOST MACROMOLECULAR SYNTHESIS BY HERPESVIRUSES DURING LYTIC INFECTION

Herpes simplex infection causes an inhibition of cell-specific RNA synthesis, DNA synthesis, and protein synthesis.

Inhibition of cellular macromolecular synthesis has been described as a common consequence following virus infection of animal cells. It generally occurs while viral macromolecular synthesis is proceeding and, therefore, requires the involvement of very specific mechanisms. Although this inhibition is not invariable, it has been reported in many different systems (for a review, see 163).

The gradual inhibition of RNA synthesis, which is observed in HSV-infected cells, concerns all classes of RNA. Functional cellular mRNA synthesis is reduced (164), Together with the amount of 4 S RNA (165). Both levels of synthesis and processing of RNA are altered, as revealed by the inhibition of 45 S precursor rRNA synthesis and the altered processing of this precursor into mature rRNA (166–169). Similar observations were reported in the case of pseudorabies-infected rabbit kidney cells.

Inhibition of host DNA synthesis has also been described as being a characteristic consequence of herpesvirus lytic infection (155,162,170–172). The analysis of host and viral DNA synthesis in infected cells is eased by the different buoyant densities of the cellular and viral DNAs. Sedimentation in CsCl gradients of extracts from infected cells labeled with radioactive DNA synthesis precursors has revealed that host DNA synthesis is progressively inhibited, whereas herpes DNA synthesis increases (Fig. 19). It has been suggested that this inhibition is due to a virus-induced protein (173). This suggestion was supported by the isolation of temperature-sensitive mutants of HSV_2 which fail to inhibit host DNA synthesis (174). Autoradiographic studies revealed that cell DNA inhibition is accompanied by extensive aggregation and displacement of chromatin to the nuclear membrane (175).

Suppression of host protein synthesis has been also reported to occur upon infection of cells by herpsviruses (171,177,178). The requirement of protein synthesis for breakdown of host polysomes after infection has been described in pseudorabies-infected cells (179). It would seem that UV-irradiated HSV_2 is still able to abolish cellular protein synthesis, which suggests an inhi-

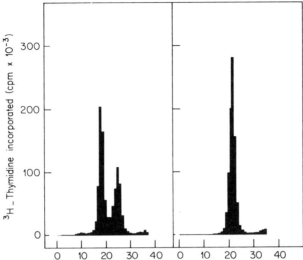

FIGURE 19. Cellular and viral DNA synthesis in HSV-$_1$-infected cells. Monolayers of BHK-21 (C13) cells were infected with HSV$_1$ (strain 17) at a multiplicity of infection of 50 pfu/cell. Viral and cellular DNA were labeled by the addition of ^{3}H thymidine (100 μCi) to the medium. Cell extracts were prepared as described by Halliburton and Timbury (174). Viral and cellular DNA (ρv = 1.725 g/cm^3 and ρc = 1.700 g/cm^3) were separated by centrifugation in CsCl (1.29 g/mL) in 50 Ti rotor (Beckman) for 72 hours at 40,000 rpm. Fractions were collected and their radioactivity measured after precipitation of DNA with 10% TCA.

bition by the particles themselves (176) rather than by newly synthesized proteins (99,171,173). A similar conclusion emerged from studies of cellular protein synthesis in uninfected cells. The rate of protein synthesis in uninfected cells was shown to remain steady, whereas it declined rapidly after infection with HSV$_2$. Under these conditions, no viral polypeptide synthesis could be detected (176).

Interestingly, the inhibition of both host DNA and protein synthesis is more efficient in HSV$_2$-infected cells than in HSV$_1$-infected cells (180–182), and mixed infection experiments have shown that HSV$_2$ infection induces the suppression of sendai virus protein synthesis, whereas it does not inhibit the synthesis of HSV_1 proteins (176). This different inhibitory potential has been used recently in a study of the level of inhibition produced by intertypic ($HSV_1 \times HSV_2$) recombinants (182). These studies led to the conclusion that one or more genes whose expression is involved in the inhibition of protein synthesis are clustered in a region of the HSV_2 genome, comprising (on the prototype arrangement) between 0.52 and 0.59 map units, together with the viral functions involved in the inhibition of DNA synthesis. Whether these inhibitions are the result of pleiotropic effects of one gene or the result of the simultaneous expression of several different viral genes remains to be established.

B. TRANSCRIPTION OF THE HERPESVIRUS GENOME

Transcription of the parental DNA molecule is the first major event following the entry of the virus in the cell during lytic infection (183–186). The most significant progress toward a better understanding of herpesvirus DNA transcription has emerged from studies performed within the past few years essentially with herpes-simplex-infected cells. Therefore, we will focus in this chapter mainly on HSV DNA transcription.

1. Herpes Simplex Virus DNA Transcription

Transcription of HSV DNA is initiated by cellular RNA polymerase II. The appearance of viral transcripts is submitted to a temporal regulation. Specific viral species have been mapped in the HSV genome. Early and late transcripts arise from noncontiguous regions of the DNA and are not clustered into separate regions of the DNA.

Viral RNA is made in the nucleus of the infected cells. Classically, the populations of RNA species synthesized before viral DNA replication occurs (see below) have been called early RNA, whereas the transcripts found in the infected cells after the onset of viral DNA synthesis have been referred to as late

RNA. The lifetime of the herpes simplex virus RNA species has been reported to be shorter than the lifetime of cellular RNA (187).

a. Cell Fractionation and Extraction of Cytoplasmic and Nuclear RNA from Herpes Simplex Virus–Infected Cells

☐ *Cell fractionation is generally achieved by lysing the infected cells in an isotomic buffer containing 0.5 to 1% of Nonidet P40 (Shell Oil Co.), followed by low-speed centrifugation (200 rpm, 10 min, 4°C) to pellet the nuclei (119,188,189). Alternatively, cells may be lysed with 2% Triton X 100 (190).*

Cytoplasmic RNA is extracted by hot phenol (50°C) in the presence of 0.5% SDS and 3 mM EDTA at pH 9.0 according to Braverman et al. (191). This method has been shown to minimize the loss of polyadenylated RNA species. After two or three subsequent extractions with 2% isoamyl alcohol in chloroform, the RNA is precipitated at –20°C with two to three volumes of absolute ethanol in the presence of 0.1 M NaCl.

Before nuclear RNA is extracted under similar conditions, nuclei are first lysed by the addition of 0.5 to 2% sodium deoxycholate (119,188) and digested with pancreatic DNAse (RNAse-free), which is often pretreated with iodoacetate to inactivate any residual RNAse activity (192). Following treatment with DNAse, the nuclear fraction may be digested with self-digested proteinase K in order to remove nuclease activity (190).

In some instances (193,194), the RNA purification schedule has been using isopycnic banding in cesium sulfate, which was reported to allow a complete separation of RNA from remaining DNA (195). However, Jones and Roizman (196) pointed out that the use of this step in herpesvirus RNA purification may result in the selective loss of high-molecular-weight nuclear RNA sequences if the purification

procedure was achieved in the absence of formaldehyde.
Under such conditions, high-molecular-weight single-
stranded RNA will aggregate at high salt concentrations
and precipitate in Cs₂SO₄ solution (197).   □

b.   Enzymatic Activities Involved in HSV Transcription.
Because intact HSV DNA strands as well as naked (deprotein-
ized) DNA duplex molecules have been reported to be infectious (198–200), it is generally assumed that the transcription of
herpesvirus DNA, at least in the early stages, does not need the
presence of a virus-specific RNA polymerase, but rather involves the participation of the host cell polymerases. Furthermore, no RNA polymerizing enzyme has ever been described in
purified virions, and the use of a-ammanitin, which selectively
inhibits host RNA polymerase II (201–203), has shown that this
enzyme is directly involved in the early transcription of HSV
DNA in infected cells (204,205). Similar conclusions were obtained by Alwine et al. (206), who showed that herpesvirus
RNA synthesis in isolated nuclei was almost completely inhibited by a-ammanitin. On the other hand, it has been reported
that viral RNA synthesis in herpes-simplex-infected nuclei is
achieved by an a-ammanitin-sensitive RNA polymerase that
can be distinguished from cellular RNA polymerase II
(207,208). The residual viral RNA synthesis observed in the
presence of a-ammanitin in infected nuclei has been attributed
to RNA polymerase III.

The observation that restricted transcription of the herpes
simplex genome occurs in the absence of protein synthesis
(194,207–210) and the studies performed with several thermosensitive herpes simplex viral mutants unable to replicate at
high temperature (DNA-negative mutants) (193) led to the conclusion that virus-specific protein synthesis was a prerequisite
for the continuation of viral DNA transcription (211). Direct
evidence has been reported suggesting that a viral protein is able
to modify the transcription of HSV genome (193,212,213).
Therefore, it would seem possible that the unmodified host

RNA polymerase II does not transcribe all the incoming genomes but rather recognizes only a few promoter sites and that a virus-coded component would be able to modify the specificity of the host polymerase or to alter the template activity of the viral DNA.

c. **Symmetrical Transcripts.** The presence of symmetrical transcripts (able to self-anneal), has been reported in herpesvirus-infected cells from several origins (214–217). The extent to which the HSV genome is transcribed symmetrically has not been resolved. Jacquemont and Roizman (217) reported that the symmetrical transcripts consist of at least two populations of RNA species, arising from 29 and 26% of the viral DNA and differing to 40-fold in molar concentrations. Only trace amounts of symmetrical transcripts are found in the cytoplasm (216), suggesting that a post-transcriptional mechanism is operating to discriminate the transcripts that will be transfered in the cytoplasm.

d. **Genetic Complexity of Early and Late RNA.** The genetic complexity of herpesvirus RNA sequences (i.e., the fraction of viral DNA homologous to viral RNA) has been estimated by hybridizing trace amounts of labeled DNA to excess unlabeled RNA in liquid phase. The analysis of the kinetics of hybrid formation under these conditions has allowed Frenkel and Roizman (218) to determine that the early RNA contains two classes of specific viral RNA which they have called abundant and scarce because the two classes differ 140-fold in concentration. These two classes were shown to be transcribed from 14% and 30% of the DNA, respectively. This was equivalent to 28 and 60% of the single-stranded RNA. Abundant early RNA contains 94% of the total virus-specific early RNA (219). The fraction of the genome which was transcribed into early RNA has been shown to comprise between 20 to 40% of the DNA (208,209,216), whereas 50% of the DNA has been found to be transcribed into late RNA (209,216,218).

Late RNA also contains two classes of abundant and scarce RNA which arise from the transcription of 19% and 29% of the DNA, respectively. These two classes differ 40-fold in concentration. Late abundant RNA was shown to contain 99% of the total virus-specific RNA made at the late time. It has also been shown that the early abundant RNA was a subset of late abundant RNA (218).

e. Properties of Nuclear and Cytoplasmic Viral RNA. It has been shown for a long time that viral transcripts appear in the nucleus as high-molecular-weight molecules sedimenting faster (40–60 S) in sucrose gradients than viral transcripts found in the cytoplasm, which sediment at about 10 to 30 S (220,221). The nuclear transcripts that sediment faster than 45 S contain sequences arising from at least 40% of the DNA. These observations suggested that the bulk of stable transcripts is synthesized in the nucleus as high-molecular-weight RNA precursors which are cleaved prior to transport in the cytoplasm.

The presence of poly-A sequence in both nuclear and cytoplasmic viral RNA was reported by Bachenheimer and Roizman (222), who suggested that the addition of poly-A tract (about 160 nucleotides long) was a post-transcriptional event. Subsequently, Silverstein et al. (219) showed that the nuclear polyadenylated viral RNA is complementary to 24% of the viral DNA, whereas cytoplasmic polyadenylated viral RNA arises from the transcription of 22% of viral DNA.

The fractionation of polyadenylated RNA from infected cells by affinity chromotography on columns of poly-U immobilized in glass fiber filters (219) allowed the isolation of three classes of polyadenylated RNA. Hybridization experiments revealed that viral RNA sequences complementary to 40% of viral DNA were present in each of the three classes of RNA which had average lengths of 30,50, and 150 adenylate residues (poly-A 30, poly-A 50, and poly-A 150 tracts).

A great proportion (57–68% of the cytoplasmic polyadenylated RNA was found to contain 155 adenylate residues, whereas 42 to 50% of the nuclear poly-A RNA contained 30 adenylate residue. The high-molecular-weight viral RNA species that sediment faster than 45 S contained almost exclusively poly-A 30 tracts. These observations suggest that adenylation is not a simple process, but rather arises by at least two separate steps. The first step would involve the adenylation of high-molecular-weight nuclear RNA (poly-A 30), whereas the second would involve the adenylation of the RNA transcripts bearing poly-A 50 and poly-A 150 tracts prior to their transport to the cytoplasm. A significant fraction of nonpolyadenylated [poly-A (−)] mRNA containing the same sequences as poly-A (+) viral RNA has been found to be bound to polyribosomes and, therefore, does not seem to require polyadenylation for transport to the cytoplasm (190).

Finally, polyribosomal viral RNA has been shown to be internally methylated and to contain a capped 5′ end (223,224) with a structure similar to that found in mRNAs of eucaryotic cells (225–227) and of some animal viruses (228–230).

Although HSV_1 mRNAs share many properties of cellular mRNAs (synthesis in nucleus, poly-A tracks, capped end), they do not seem to be extensively spliced. (See below.)

2. Temporal Regulation of Transcription and Location of the DNA Sequences Transcribed during Infection on the Genome of Herpes Simplex Virus

a. Temporal Regulation of Transcription. Analysis of the genetic complexity of viral RNA sequences found in infected cells treated with inhibitors of protein synthesis (cycloheximide or emetine) or with inhibitors of DNA synthesis [phosphonoacetic acid (PAA)] as compared to those found in untreated infected cells has brought evidence that transcription of the HSV genome is temporally regulated. These studies suggest the existence of three phases in the transcription program (196).

During phase 1, the transcription of HSV genome would involve host proteins (i.e., RNA polymerase II), either alone or in combination with virus-specified proteins. In phase 2, the transcription would be mediated by proteins synthesized after infection, but before viral DNA replication has been initiated, and in phase 3, transcription would be coupled to initiation of replication. The existence of three cycles of transcription would be coupled to initiation of replication. The existence of three cycles of transcription is in agreement with the temporal regulation governing protein synthesis in HSV-infected cells (see below, synthesis of α, β and γ polypeptides). The use of cycloheximide has allowed the determination that the abundant class of early RNA which is homologous to 20 to 25% of the viral DNA is composed of two subclasses called α RNA, or immediate early (IE) RNA, and β RNA, or delayed early (DE) RNA. In the presence of cycloheximide only the α (or IE) RNA is found to accumulate in the cytoplasm of the herpesvirus-infected cells (119,194,208–210,216,218), although the transcripts which accumulate in the nucleus of such cells are similar to those found associated with the polyribosomes late in infection in untreated cells (216). Furthermore, upon withdrawal of the drug, a new wave of RNA synthesis is necessary if the β and late RNA species in the cytoplasm are to be found. In each of the three phases defined above, the genetic complexity of the nuclear viral RNA has been shown to be greater than that of cytoplasmic RNAs (196) (Table II). It seems, therefore, that post-transcriptional events are needed for transport of RNA species to the cytoplasm (119). Similar conclusions have been obtained in previous studies of the genetic complexity of late nuclear and cytoplasmic RNAs (216,218,231). Because there are three distinct classes of viral mRNA species being described in HSV-infected cells (IE, DE, and late RNA), it was of interest to determine whether these transcripts arose from clustered sequences on the genome, as in the case of Polyomavirus (see chapter I), or from noncontiguous regions, as in the case of adenovirus transcription.

TABLE II. Fraction of Viral DNA Homologous to Viral RNA from Nucleus and Cytoplasm of Infected Cells[a]

Phase No.	Growth Conditions	Nuclear RNA	Cytoplasmic RNA
1	Absence of protein synthesis (+ cycloheximide)	0.33	0.12
2	Absence of viral DNA synthesis (+ PAA)	0.39	0.26
3	Untreated cells (8–14 hours postinfection)	>0.50	0.41

[a]Data from Jones and Roizman (196).

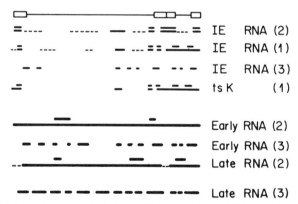

FIGURE 20. Mapping of viral-specific cytoplasmic RNA species on HSV₁ genome. Dotted lines indicate low levels of hybridization, a single continuous line indicates clearly detectable hybridization, and a double line is for relatively abundant hybridization. Data from Watson and Clements (193), Clements et al. (194), and Jones and Roizman (196).

b. Location of the Transcripts on HSV Genome

i. EARLY RNA. The location of viral transcripts synthesized in the presence of cycloheximide has been determined by hybridization of radiolabeled viral RNA to denatured restriction endonuclease fragments that have been separated by electrophoresis on agarose gels and transferred to a nitrocellulose membrane according to the technique introduced by Southern.

The mapping of cytoplasmic a (IE) RNA has shown that these transcripts do not arise from clustered regions on the genome but, rather, hybridize to scattered restricted portions of the HSV genome and especially to the repeats which flank the S region. Hybridization also occurs with unique DNA fragments from the L region, with fragments from the repeats flanking the L region, and with DNA fragments which originate from the ends of the L region (Fig. 20) (119,194,196). Cytoplasmic RNA isolated from infected cells incubated during 7 hours in the presence of cycloheximide was found to be homologous to 12% of HSV_1 DNA (196) as previously reported (216).

If one assumes that more than one of the four possible isomers is infectious, the fact that a (IE) RNA templates are found on both L and S components implies that RNA synthesis proceeds on different strands in at least two of the four DNA arrangements (196).

Conflicting results have been reported concerning the nature of the a (IE) RNA accumulated in the nucleus of infected cells upon incubation in the presence of cycloheximide. Jones and Roizman (196) reported that transcripts homologous to DNA segments located between 0.18 and 0.53 map units on the genome have been found in the nuclear a (IE) of infected cells incubated in the presence of cycloheximide or emetine, whereas they were not detected among cytoplasmic a (IE) RNA. On the contrary, Clements et al. (194) did not observe any difference between the hybridization patterns of cytoplasmic and nuclear a

(IE) RNA. In addition to the distinct host cell systems used, different technical procedures may be encountered for the observed discrepancies (196).The analysis of nuclear and cytoplasmic transcripts synthesized in DNA-negative thermosensitive mutants and incubated at the restrictive temperature revealed that one of them (tsK of Glasgow strain) exhibits a transcription pattern similar to that obtained in the presence of cycloheximide (193) and that the temperature-sensitive lesion in this mutant lies in an immediate early polypeptide which has been shown to be directly responsible for the induction of new viral transcripts including mRNA coding for pyrimidine deoxyribonucleoside kinase (dPyk) (193,212,213). These observations are in agreement with the previous studies of Leung (232), who showed that a protein factor encoded by a (IE) RNA is involved in the expression of dPyk mRNA in productively infected cells. The viral polypeptide affected by tsK mutation is VMW 175,000 (also called ICP$_4$) (233).

ii. PROPERTIES OF a (IE)RNA SPECIES. Purification of polyadenylated immediate early transcripts by electrophoresis on denaturing agarose gels containing methyl mercuric hydroxide (CH$_3$HgOH) has allowed the isolation of distinct virus-specific RNA populations. Five predominant IE RNA species designated IE mRNA-1 through IE mRNA-5 have been mapped on the genome (234,235). IE mRNA-1 (3.0 Kb) and IE mRNA-3 (4.7 Kb) map wholly within the reiterated sequences TR$_L$/JR$_L$ and TR$_S$/IR$_S$, respectively, whereas IE mRNA-2 (2.0 Kb) was found to map in U$_L$. IE mRNA-4 and IE mRNA-5 (both 2.0 Kb) were mapped at or close to the junctions of U$_S$ and TR$_S$/IR$_{S^-}$. Fine mapping performed by S$_1$ nuclease depletion and R loop analysis showed that these two mRNAs contain a 5' terminal sequence mapping in TR$_S$/IR$_S$, spliced to non colinear transcripts containing both reiterated and unique sequences. *In vitro* translation of these mRNA species showed that 4.7 and 3.0 Kb mRNAs specify the synthesis of polypeptides having a molecular weight of 17,500 and 110,000 daltons, respectively, whereas the 2.0 Kb mRNAs specify the synthesis of three different polypeptides of molecular weights of 68,000, 63,000, and 12,000 daltons. IE mRNA-6 and IE mRNA-5 specify virus-specific polypeptide 68,000 and 12,000. On the other hand, Millette and Talley-Brown (236) also reported the isolation of individual IE mRNA species whose sizes were found to be 5.2, 3.8, 2.7 and 1.8 Kb. *In vitro* translation of the three first species led to the synthesis of polypeptides with molecular weights of 165,000, 145,000, and 123,000 daltons. The fourth mRNA species encoded three polypeptides with molecular weights of 86,000, 71,000, and 55,000. These results suggest that extensive intervening sequences are not involved in the expression of these genes. Cytoplasmic a RNAs obtained in the presence of the inhibitors anisomycin, emetine, and cyloheximide have been translated *in vitro*. In all cases, the major a polypeptides ICP_4, ICP_0, ICP_{22}, and ICP_{27} were found among the translation products. This observation suggests that a transcription is probably mediate by a virion-associated factor. Evidence has also been obtained that shows that ICP_4, 0, 22, and 27 mRNAs are transcribed from separate promotors and that ICP_4 0, and 27 mRNAs are not spliced near their 5' termini (S. MacKem, International Workshop on Herpes Viruses, July 27–31, 1981, Bologna, Italy).

A very elegant approach to the study of a gene expression in HSV_2-1 infected cells has been reported by Post et al. (237). Their investigation relied on the fusion of the 5' end of an a gene (ICP_4) to the structural gene sequence of the β gene coding for thymidine kinase whose enzymatic activity can be early assayed. The recombinant was used to convert LtK⁻ cells to tK⁺ phenotype or was inserted into the viral genome.

In cells infected with the recombinant virus, the thymidine kinase gene was expressed as an a gene. In other words, the thymidine kinase gene was expressed in the absence of prior infected cell protein synthesis.

These studies showed that optimal a gene expression requires the expression of one or several new a genes (such as one or more virion components).

Furthermore, this system not only offers the possibility of finding a better approach to the study of a gene expression in HSV_1-infected cells, but it also appears to be an excellent system for the study of the expression of other eucaryotic promoters.

iii.   LOCATION OF TRANSCRIPTS FOUND IN PHASE 2.   The hybridization patterns of immediate early and early RNA species have been found to differ essentially because of the low abundance of early transcripts that hybridize to repeats flanking the short unique region and the distribution throughout the genome of DNA fragments complementary to these early RNA species (194).

Jones and Roizman (196) also showed that the increase in genetic complexity which has been reported to occur during phase 2 of the transcriptional program (see above) is due to the accumulation of transcripts arising from the region bound by 0.18 and 0.42 to 1.00 (see Fig. 21). Location of early transcripts by RNA displacement loop (R loop) mapping on the HSV genome confirmed that the early gene regions are not completely contiguous (238). To date three β mRNAs have been mapped rigorously: the 1.5 Kb mRNA coding for thymidine kinase at about 0.3 map unit in the long unique region; a 5.2 Kb mRNA coding for a 140,000 dalton polypeptide at about 0.57; and 1.8 Kb mRNA coding for a 64,000 dalton polypeptide at about 0.62 (247).

The size distribution of HSV_1 mRNA accumulated both in the nucleus and the cytoplasm of infected cells before viral DNA synthesis is initiated (early RNA) has been analyzed by

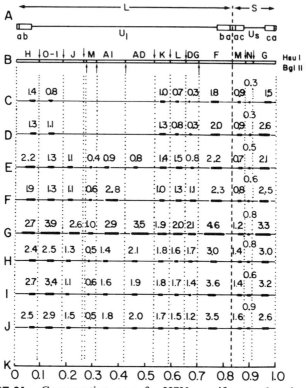

FIGURE 21. Comparative maps for HSV₁ specific cytoplasmic and nuclear RNA. The bars represent the relative abundance and location of cytoplasmic RNA transcripts accumulated in the presence of (C) cycloheximide, (D) emetine, (E) canavanine, (F) PAA, and (G) without inhibitor after 8 hours infection and of the corresponding nuclear RNA accumulated in the presence of (H) cycloheximide, (I) emetine, and (J) canavanine. The arrows indicate the restriction sites for HSV I and Bgl II restriction nucleases (B). Data from Jones and Roizman (196).

electrophoresis on denaturing (CH₃HgOH) agarose gels (Fig. 22). Under these conditions, migration has been shown to be linearly related to the logarithm of RNA size (239). The results obtained (240) showed that the size distribution of viral RNA in

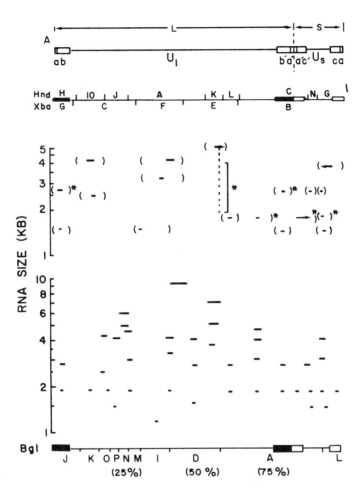

FIGURE 22. Size of early and late HSV₁ transcripts. Electrophoresis and hybridizations have been performed as described in the text. The length of the lines is proportional to that of the corresponding DNA. The locations of Hind III, Xba I, and Bgl III restriction fragments are shown for the P orientation of HSV DNA (a) early RNA, (b) late RNA. Data from Holland et al. (240) and Anderson et al. (245).

the nucleus and on polyribosomes was similar, suggesting that the promoters of the mRNA species are near the coding sequences for the considered RNA. Hybridization of size-fractioned RNA species to denatured restriction endonuclease DNA fragments has allowed the mapping of 16 viral mRNA species on the HSV genome. Although these species account for a large proportion of the genetic complexity of the abundant early mRNA, the establishment of more detailed restriction maps will certainly allow the location of additional mRNA species on the viral genome. A correlation between polypeptide size and mRNA coding capacity has been possible in nine cases. Interesting is the observation that the smallest mRNA species (1.5Kb) characterized in this study have significantly more coding capacity than that required for the synthesis of the smallest polypeptides (30,000 daltons) and may, therefore, contain significant untranslated sequences (240).

iv. LOCATION OF LATE TRANSCRIPTS. The third phase of the transcription program was defined as being coupled to viral DNA synthesis (196). Several studies on protein synthesis (241–244) suggested that viral DNA replication might be involved in the regulation of herpes DNA transcription. When, at late times, the viral DNA synthesis is reaching its maximum rate, the great majority of the early RNA transcripts are still synthesized (194,208), and new transcripts (γ RNA) arising from 15 to 20% of the viral genome are found on polyribosomes. At least a portion of these γ RNA species have been shown to be present among the early transcripts. However, they are found at a concentration 10- to 20-fold lower than α and β mRNA species (190,210,216). The increase of genetic complexity of the cytoplasmic RNA during phase 3 has been shown to result from the accumulation of transcripts which arise from the region 0.42 to 0.53 map units on the genome (which was not transcribed in phases 1 and 2) as well as from other sequences arising from other regions of the viral genome (196).

Polyadenylated mRNA species which accumulate at late times after infection and arise from about 45% of the viral DNA have been localized on the HSV genome, either by purifying viral mRNA on HSV DNA restriction fragments bound to cellulose and sizing of the hybridized species on denaturing (CH_3HgOH) agarose gels or by first fractionating the mRNA species on agarose gels and then hybridizing the different fractions to blots of HSV DNA restriction fragments (245). These studies allowed construction of a preliminary transcription map of the HSV_1 genome. The map obtained (Fig. 22) revealed that two areas of the long unique region are complementary to a large number of HSV transcripts, and that large mRNA species (greater than 5 Kb) were found complementary to fragments that originate only from the middle of the long unique region. The sizing of the mRNA species synthesized at a late time in infection showed that the largest discrete species were about 8 to 9 Kb in size, and the smallest were found to be not less than 1.2 Kb. Such mRNA size corresponds to a coding capacity twice that required to code for the smallest viral-induced polypeptide whose size has been estimated at about 15,000 daltons (see Section IV). The distribution of the mRNA sizes also reveals that there appear to be some favored sizes, such as 1.5 to 1.5 Kb, 1.8 Kb, 2.2 Kb, 5.4 Kb, and 6 Kb, with a predominant proportion of 0.8 to 0.9 Kb species. Again, this is larger than the size required to code for polypeptides of molecular weights less than 60,000 daltons, which were shown to account for half the identified viral polypeptides (see Section IV). These observations and the little amount of the post-translational processing occurring in HSV-infected cells (246) led Anderson et al. (245) to suggest that many molecules may contain a significant amount of untranslated sequences. More recently, Anderson et al. (247) mapped several γ mRNAs on the prototype arrangement of HSV_1 genome and concluded that no internal introns are found in the coding sequences of these mRNAs. Also, the 5′ end of several take mRNAs that are coded on opposite DNA

strands map very close to each other. The use of two recombinant DNA clones carrying either the EcoRl G fragment (0.190–
0.30 map units) or the Hind III J fragment (0.181–0.259 map
units) allowed Costa et al. (248) to characterize in great detail a
major late HSV$_1$ mRNA (6 Kb) mapping in the large unique
region of the HSV$_1$ genome. The 3' end of this mRNA lies on
the left of 0.231 map position and is transcribed from right toward left. The 3' end of this mRNA is partially colinear with a
1.5 Kb mRNA encoding a 35,000 dalton polypeptide. No in-
tron was detected in this 6 Kb mRNA, which codes for a
155,000 dalton polypeptide.

3. *In vitro* Transcription of the HSV Genome

Purified *Escherichia coli* RNA polymerase has been used to direct *in vitro* transcription of HSV DNA (249,250). The transcripts synthesize after a brief incubation time (30 min. at 37°C)
have been analyzed by centrifugation and blotting to fragmented HSV DNA. The results obtained revealed the presence
of two classes of RNA species (less than 2 Kb and between 2
and 4 Kb in length), both of which hybridized to all the fragments but one (K fragment) obtained after EcoRl digestion. The
size of the transcripts obtained in these experiments was not
very different from the size of the major mRNA species described by Anderson et al. (245). When *in vitro* transcription
with *E. coli* polymerase is performed for a longer incubation
time (3 hours at 37°C), the RNA species synthesized are found
to be complementary to all the HSV DNA fragments obtained
after digestion with either Hpa I or Hind III restriction nuclease
(249).

   The use of purified eucaryotic RNA polymerase II (isolated
from calf thymus) to direct *in vitro* synthesis of herpes-specific
RNA leads to the isolation of RNA species whose size does not

exceed 2.5 Kb in length and that hybridize to all the fragments obtained after EcoRl digestion (Fig. 23). Both *E. coli* and calf thymus RNA polymerases transcribe the HSV genome on both strands as shown by the high proportion of symmetrical transcripts observed among *in vitro* transcription products (250).

In vitro transcription of whole HSV_1 DNA and cloned HSV_1 DNA fragments by RNA polymerase II purified from Hep 2 cells has been reported by T. W. Beck and R. L. Millette (Abstract of International Workshop on Herpes Viruses, July 27–31, 1981, Bologna, Italy). Transcription of whole HSV_1 DNA yielded high-molecular-weight RNA species (4 S to greater than 28 S) whose 12% appeared to be self-complementation. In these experiments, RNA polymerase II was found to transcribe essentially all regions of the viral genome although preferential transcription of IE (*a*) genes was reported.

The results obtained along *in vitro* transcription of the cloned Bam H1 Q fragment coding for thymidine kinase and mapping at coordinates 0.29 to 0.31 raised the possibility that four promotors may be used by RNA polymerase II in this region (Fig. 24).

C. TRANSCRIPTION OF OTHER *HERPESVIRUS* DNAS

1. Pseudorabies Virus and *Herpesvirus saimiri*

A detailed analysis of the pseudorabies virus-transcription program has been undertaken recently. The results available thus far show that in the absence of protein synthesis only 12% of the viral genome is transcribed, giving rise to *a* (IE) RNA species that hybridize only to the ends of the inverted repeated regions of the genome (251). Previous studies (252) had shown that 25% of the early transcripts made in untreated cells could

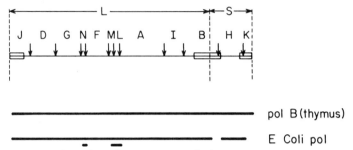

FIGURE 23. Location of *in vitro* transcription products on HSV genome. The continuous line indicates the uniform level of transcription obtained with calf thymus polymerase B, whereas double lines indicate preferential sites of transcription for *Escherichia coli* polymerase. The arrows indicate the restriction sites for EcoR1 restriction nuclease. Data from Chenciner (250).

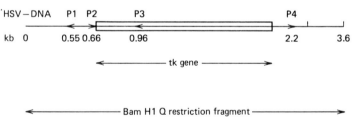

FIGURE 24. Model for *in vitro* transcription in HSV-₁ Bam H1 Q DNA. This model shows the different promoters found in this region. The RNAs initiated from promoters P_3 and P_1 would have common 3′ sequences but different 5′ sites, whereas RNAs intiated at P_2 and P_3 would overlap for about 240 bases from their 5′ ends. The P_4 site correlates with a TATTAA sequence. P_1 and P_3 give leftward transcripts, P_2 gives rightward transcripts with respect to the thymidine kinase (tk) gene sequence map. Data from T.W. Beck and R.L. Millette. (International Workshop on Herpes Viruses July 27–31, 1981, Bologna, Italy.)

110

be detected in the presence of cycloheximide. The only sequences found up to 30 min. in untreated cells after infection are a (IE) RNA species (251). There is a decrease thereafter in the rate of a (IE) RNA synthesis which is concomitant with an increase in the transcription of early genes. A similar switch between a (IE) RNA and early RNA is observed upon removal of cycloheximide from infected cells grown in the presence of this drug. Blotting patterns of late viral RNA transcripts showed that there is no selective and extensive degradation of RNA sequences at late times during infection.

The isolation of the polyadenylated RNA species purified from cells of owl monkey kidney (OMK) infected by *H. saimiri* has allowed hybridization studies with H (high G + C) and L (low G + C) DNAs (see structure of *H. saimiri* in Section II). Upon infection, most of the polyadenylated RNA synthesized has been shown to consist of viral RNA (253). Hybridization studies have revealed that less than 1% of the poly-A RNA hybridized to H DNA, whereas the hybridization of poly-A RNA to L DNA accounts for all the virus-specific poly-A RNA. It is interesting, therefore, that a very small fraction of the viral DNA (1%) arises from the repetitive H DNA regions which represent up to 30% of the total *H. saimiri* genome (152).

2. Epstein-Barr Virus

The reader interested in the transcription of Epstein-Barr virus should consult the excellent review of Kieff et al. (139). Salient features of our knowledge in this field are as follows.

In restringently infected cells, in which a partial restricted expression is observed, a substantial amount of viral DNA is transcribed into stable RNA (254,255). These RNA species have been found to be complementary to 18% of EBV DNA isolated from Namalwa and Kurgans cells. Some RNA species synthesized in such cells selectively accumulate on polyribosomes (254). Both the viral transcripts from restringently infected cells and from Burkitt's tumor tissue have been located on the map

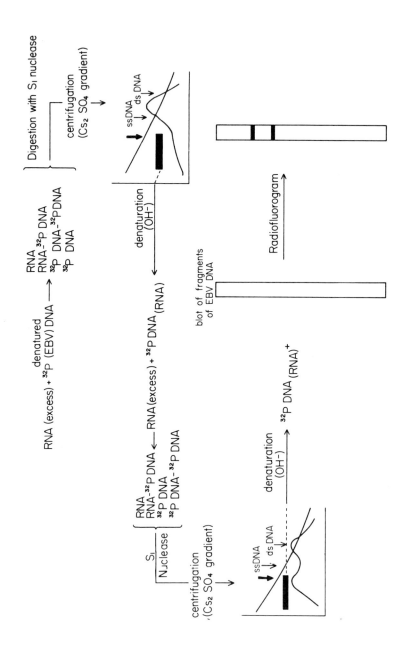

of EBV genome either by hybridizing poly-A(+) and poly-A(-) RNA species to purified fragments of EBV DNA, or by identifying, on the EBV genome, the fragments that contain sequences homologous to poly-A(+) and poly-A(-) RNAs (Fig. 25) (256). The location of EBV RNA transcripts on the viral genome is shown in Fig. 26.

Cultures containing productively infected cells have been reported to contain additional RNA species, the total viral RNA found in productively infected cells being complementary to at least 45% of the viral genome (255,257).

Enhanced processing of viral RNA could be responsible for the transition from restringent to abortive infection (139).

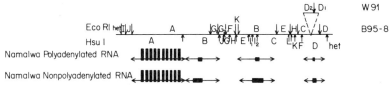

FIGURE 26. Map of the HsuI and EcoRI restriction endonuclease fragments of EBV (B95-8) and (W-91) DNA with diagrammatic representation of the location of DNA sequences encoding stable RNAs in Namalwa cells. The width of the bars indicates the length of DNA which is homologous to stable DNA, within the restriction fragments. Data from Powell et al. (256).

FIGURE 25. (Opposite) Identification of restriction endonucleases fragments of EBV DNA homologous to viral RNA in Namalwa cells. A preparation of ^{32}P-labeled EBV DNA was first hybridized to poly-A(+) or poly-A(-) RNA and subsequently incubated with S$_1$ nuclease. The mixture was then centrifuged in a cesium sulfate gradient in order to select for DNA-RNA hybrids. The positions of single stranded and double stranded DNA were indicated by ^{3}H *Klebsiella-pneumoniae* DNA added to the sample. After partial denaturation of the DNA-RNA hybrids contained in fractions having densities from 1.5 to 1.62 g/ml, a second hybridization with poly-A(+) or poly-A(-) RNA and a second gradient purification were performed. The resulting ^{32}P DNA was hybridized to a nitrocellulose blot of EBV-digested DNA. The location and relative amounts of each EBV DNA fragment on the blot were measured independently after hybridization with purified total ^{32}P EBV DNA.

D.   REPLICATION OF *HERPESVIRUS* DNA

Replication of herpesvirus DNA involved virus-specified proteins and is semiconservative. The origin of replication is located in the S arm portion of the joint region. A rolling circle mechanism has been put forward to account for the known features of herpes DNA replication.

Most of the available information concerning herpesvirus DNA synthesis has emerged from studies performed with herpes simplex virus and pseudorabies virus-infected cells.

Histochemical and biochemical studies have shown that herpes DNA synthesis takes place in the nucleus of the infected cells (258–260,162,5).

1.   Requirements for *Herpesvirus* DNA Replication

The synthesis of herpesvirus DNA is generally independent of the growth cycle of the cells (261). However, murine and human cytomegalovirus DNA replications have been reported to be dependent upon the S cell cycle phase of cultivated mouse fibroblasts (262,263). Conflicting observations have been reported as to the need of cellular DNA replication for EAV DNA synthesis to proceed (264–266). Viral DNA is generally detected within a few hours after infection, the bulk of synthesis taking place between 4 and 7 hours after virus penetration (99,162,267). Thereafter, an irreversible decline of viral DNA synthesis is observed (162) (Fig. 27). The synthesis of herpes DNA is dependent upon concomitant protein synthesis. The addition of puromycin to infected cells at any time between 3 to 4 hours after infection leads to the inhibition of herpes DNA synthesis (7,268).

Upon infection, the levels of several enzymes involved in the metabolism of the nucleic acids increase. This is due to the synthesis of several virus-coded enzymes, which can be distinguished from their cellular counterparts on the basis of their particular biochemical properties. Among these enzymes are

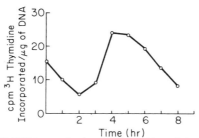

FIGURE 27.   HSV DNA synthesis as a function of time after infection. Bulk of HSV DNA synthesis, measured by incorporation of ^{3}H thymidine, is shown to occur between 4 and 7 hours after virus penetration. Data from Roizman (162).

DNases (184,269–275), thymidine kinase (276–279), deoxycytidine kinase (280), deoxycytidine desaminase (281,282) and ribonucleotide reductase (283–285). New DNA polymerase activities have been described in cells infected with herpes simplex virus (280,286–290), pseudorabies virus (291), Marek's disease (292), human cytomegalovirus (293,294), bovine rhinotracheitis virus (295), and equine herpesvirus (296–298).

The use of phosphonoacetic acid (PAA), an inhibitor of herpesvirus DNA polymerase (297,299–311) (Fig. 28), has allowed the isolation of resistant mutants (230,312,313) which have been used to locate on the viral genome map the corresponding cistrons (see Section V). Mao et al. first showed that PAA is able to inhibit herpes-induced DNA polymerase activity in vitro; subsequently Hay and Subak-Sharpe (312) showed that the DNA polymerase induced by PAA-resistant mutants is more resistant to PAA *in vitro* than that of wild-type parental virus strain.

More recently, evidence has been reported to suggest that a single HSV$_1$ gene codes for a multifunctional enzyme with thymidilate kinase activity as well as thymidine kinase and deoxycytidine kinase activity (314).

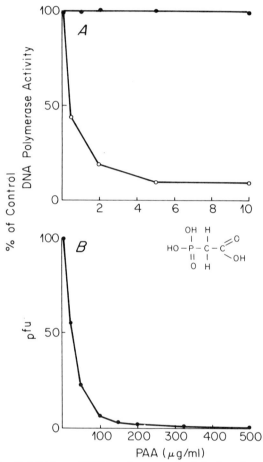

FIGURE 28. Inhibition of DNA polymerase activity by phosphonoacetic acid (PAA). The structure of PAA is represented. The inhibition of herpes DNA replication is shown by the decrease of infectious particles obtained after infection of BHK cells by frog virus 3 (311). The data are expressed as percentages of the yield obtained in the absence of drug (B). The *in vitro* inhibition of HSV DNA polymerase (A) is expressed as a percentage of the activity obtained without PAA. Filled dots represent the activity of the DNA polymerase in a PAA-resistant strain. (A) Data from Elliott et al. (311) and (B)from Honess and Watson (239).

In addition to DNA polymerase mutants, several other mutants unable to be replicated at high temperature [thermosensitive DNA(-) mutants] have been used to characterize further the other functions involved in herpesvirus DNA replication. The results obtained show that at least three viral proteins are continuously required for the synthesis of viral DNA, one of them being the viral-induced polymerase (315–317).

Studies performed with nonpermissive neuronal infected cells showed that the induction of viral DNA polymerase synthesis is not sufficient to allow DNA replication to proceed (318). A similar block in replication has been put forward to explain the nonproductivity of oncornavirus-transformed cells superinfected by herpes simplex virus (319). These kinds of studies might well define other conditions involved directly or indirectly in herpes DNA replication. Interesting in this matter are the observations that cells containing high levels of SV40 T antigen are more resistant to HSV than cells with lower levels of T antigen (320) and that replication of herpes simplex virus may be altered in simian adenovirus-transformed cells (321). Finally, integration of herpes simplex type 1 DNA with host chromosomal DNA during productive infection has also been reported (322,323).

2. Physical Properties of Newly Synthesized *Herpesvirus* DNA

A single viral particle can be sufficient to initiate the infection process (324). Newly synthesized herpesvirus DNA is made of fragments (156,158) to which RNA sequences are covalently linked (325). These fragments form an intracellular pool from which DNA is withdrawn at random to be integrated into viral particles (86,326,327).

Herpesvirus DNA replication is semiconservative, both strands of the molecule being able to act as a template (86), and it may involve a repair-type mechanism (334). Most of the single-stranded interruptions in mature pseudorabies viral DNA were shown to be ligated prior to replication of the DNA (328). Sedimentation of newly synthesized HSV_1 DNA in alkaline gradients showed that nascent DNA is found as small fragments elongated later. Thus, DNA labeled for 3 min. is found to be 0.5 to 1.0×10^6 daltons long, and this size increases with duration of labeling time (2).

The rate of fork movement for pseudorabies DNA elongation has been estimated to be approximately 1 μ/min at 37°C in infected cells (329), and electron microscopic data suggest that replication of HSV_1 DNA is initiated close to one end of the molecule and continues bidirectionally (330). An HSV DNA-protein complex, active in DNA synthesis *in vitro,* has been purified from nuclear and chromatin preparations from infected cells (331).

Based on the assumption (332) that during a short pulse of a radioactive precursor to the infected cells the replicating DNA molecules will be predominantly labeled at the origin, whereas the completed and mature molecules will be labeled at the termini, Hirsch et al. (333) studied the distribution of labeling in the replicating HSV DNA. The kinetics of radioisotope incorporation into specific fragments of the replicating molecule were used to locate the origin of DNA replication. Replicating DNA was labeled with [^{3}H]dT, mixed with purified ^{32}P-labeled DNA, and digested with restriction endonucleases. The comparison of the ^{3}H/^{32}P ratio for the different fragments generated after endonuclease digestion allowed the location of the origin of replication in the joint region. The relative specific radioactivities of the adjacent fragments suggested that the origin of DNA synthesis may be located in the S arm portion of the joint region (Hind III, fragment N) rather than the L arm portion of the joint (Fig. 29 and Table III).

Sedimentation of newly synthesized HSV DNA present in infected cells revealed the presence of molecules smaller than unit

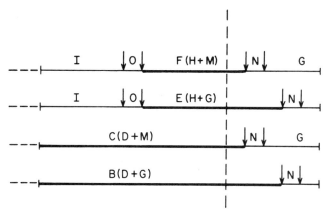

FIGURE 29. Enlargement of the joint region of HSV-$_1$ genome showing the restriction fragments generated with Hind III endonuclease. Heavy lines represent the joined fragments generated by inversions of the long and short unique regions in the four different arrangements of HSV$_1$ DNA. Data from Skare and Summers (116).

size, molecules of unit size, and molecules longer than unit size (333). Newly synthesized pseudorabies virus DNA also behaves as molecules with high S values when sedimented in neutral sucrose gradients (Fig. 30), whereas newly synthesized cellular DNA does not behave in this manner (324) and migrates to a different location that mature DNA upon electrophoresis in acrylamide gels (334). Electron microscopy studies have revealed the existence of a single-stranded region in the replicating DNA, and the presence of circular DNA molecules for both HSV and pseudorabies virus (335,336).

Viral DNA molecules with a partial circular conformation were also observed. Examination of purified intracellular pseudorabies virus DNA have revealed the presence of large tangles of DNA with densely aggregated areas near their center (336). It was suggested, therefore, that the high S value obtained in sedimentation studies is due to the presence of concatemeric forms of viral DNA. The presence of multiunit concatemers in nuclei of infected cells containing HSV DNA was also reported (337).

TABLE III. Relative Labeling of Hind III Fragments from HSV-[1] DNA Replicating Molecules Labeled with ³H Thymidine[a]

| Hind III Fragments | Molar Ratio | [³H] to [³²p] ratio | |
		Continuous Labeling for 12 hours p.i.	Pulse Labeling 15 Min. (at 12 hours p.i.)
A	1.0	1.17	0.33
B	0.25		
C	0.25	1.12	*0.68*
D	0.5	1.11	0.41
E	0.25	1.08	*0.81*
F	0.25	1.00	*1.00*
G	0.5	0.82	0.39
H	0.5	0.94	0.47
I	1.0	1.05	0.56
J	1.0	0.83	0.55
K	1.0	0.91	0.63
L	1.0	0.93	0.65
M	0.5	0.95	0.51
N	1.0	1.05	0.68
O	1.0	0.98	0.58

[a] See text for discussion of data. Data from Hirsch et al. (333).

Recent studies (338) have shown that viral DNA that accumulates in the nuclei of HSV-infected cells consists of head-to-tail concatemers which arise from the replication of the DNA by a rolling circle mechanism such as that described by Gilbert and Dressler (339). It was shown previously that all four DNA arrangements of the HSV genome were found in equimolar concentration in the viral progeny (108,109,117,118,340) and, therefore, must have arisen in the course of viral replication as a consequence of some obligatory event. The model proposed by Jacob et al. (338) presents the generation of all four insomers from one initial arrangement on the basis of excision and repair of unit length DNA from the head-to-tail concatemers (Figs. 31–32). The model presented is consistent with many of the features of viral DNA structure and synthesis.

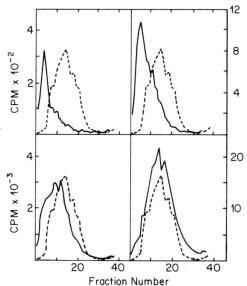

FIGURE 30. Sedimentation of newly synthesized pseudorabies DNA in alkaline sucrose gradients. Newly synthesized DNA was labeled by the addition of ³H thymidine (50 μCi/ml) to the medium at 4.5 hours postinfection. Cells were harvested at 5 min (top left), 10 min (top right), 30 min (bottom left), and 120 min (bottom right) and analysed on alkaline sucrose gradients (5–20%). The position of [¹⁴C] thymidine-labeled mature *Pr* viral DNA is shown with dotted lines.

IV.   HERPES POLYPEPTIDE SYNTHESIS IN INFECTED CELLS

Herpesvirus polypeptides are synthesized in the cytoplasm of the infected cells. Their synthesis is coordinately regulated and sequentially ordered in a cascade fashion. The location of the corresponding templates on the HSV genome reveals that all the regions of the DNA molecule, including the repetitive sequences, are expressed.

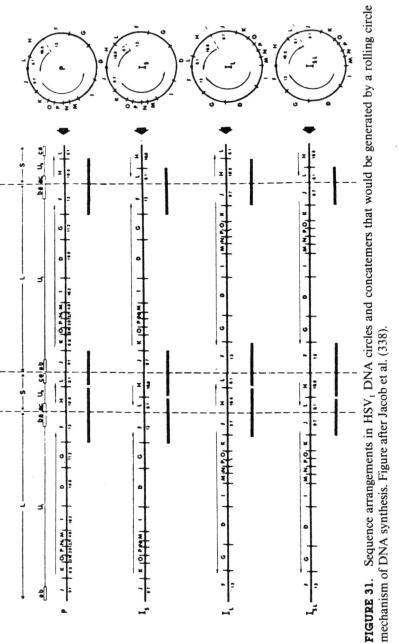

FIGURE 31. Sequence arrangements in HSV, DNA circles and concatemers that would be generated by a rolling circle mechanism of DNA synthesis. Figure after Jacob et al. (338).

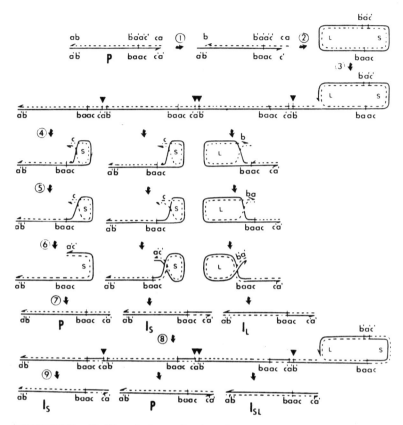

FIGURE 32. Rolling circle mechanism for the replication of HSV₁ DNA. After infection, HSV DNA is digested by a processive exonuclease, exposing cohesive simple-stranded ends (step 1). The DNA then circularizes (step 2). The junction bac differs from the internal b'a'a'c' junction. The DNA is then replicated undirectionally by a rolling circle mechanism (step 3). The resulting concatemer is split at the bac junction (either to the left or to the right of a sequence) to generate unit size DNA. Figure after Jacob et al. (338).

Viral protein synthesis in herpesvirus-infected cells takes place exclusively in the cytoplasm (178,246,341,342). After an initial decline of the amount of polyribosomes specifying host proteins, a period of increased rate of protein synthesis is observed, as measured by the incorporation of labeled amino acids into newly synthesized polypeptides. Several lines of evidence show that these polypeptides are virus-specified. The kinetics of protein synthesis, measured by quantitation of labeled polypeptides resolved by polyacrylamide gel electrophoresis, reveals that the host protein synthesis is greatly reduced, whereas new labeled polypeptide species appear at increasing rates. These new species have a molecular weight similar to that of the polypeptides found in purified virions and may be specifically precipitated by antisera prepared against viral proteins (343–345). However, some polypeptides, whose rate of synthesis increases in infected cells, have sufficiently different migrations to be classified as nonstructural polypeptides (47,246).

More than 50 virus-specified polypeptides have been described in herpes-simplex-infected cells (47,180,246,346) with molecular weights ranging from 11,000 daltons to 273,000 daltons. Several herpesvirus-induced polypeptides are submitted to a post-translational processing, such as glycosylation (347–355), sulfation, (356–358) and phosphorylation (181,353–355). Therefore, it may be ambiguous to designate the viral polypeptides by their apparent molecular weight. However, the lack of uniformity in the domination of these polypeptides, which results from different purification schemes, variety of gel systems, or different virus strains, renders very difficult the comparison of the results obtained by several groups. The molecular weight of the structural, nonstructural, phosphorylated, and glycosylated viral polypeptides detected in HSV_1 (17 mp) infected hamster cells are reported in Tables IV and V. A comparison of viral proteins induced by different herpesviruses is presented in Tables VI and VII.

TABLE IV. HSV-¹ (17 mp) Specified Polypeptides in BHK-21 (C13) Infected Cells[a]

Structural	Nonstructural
273	
248	
239	
175	
155	
145	
136	
129	
122	
117	114
110	103
100	
87/85	
82/81	
74	
	68
67	
65/64	60
57	54
51	45
43	41
	40
	39
	38
37'(37.5/37)	34
	30
28	
	27
	23
	22
21	
	19
	17.5
	17
16.5	
	16
15	
14	
13	
12.5	
11	

[a] Data from Marsden et al. (47).

TABLE V.   Viral Phosphoproteins and
Glycoproteins   Induced   in   BHK-21
(C13) Cells Infected with HSV-1
(17 mp)

Phosphoproteins	Glycoproteins
175	122
136	117
117	110
87	82,81
85	73'
82	68
68	60
65' (65)	58
58	57
65' (55)	51
45	
38	
36	
32	
28	
21	

A.   TEMPORAL REGULATION OF HERPES SIMPLEX PROTEIN SYNTHESIS

The induction of viral polypeptide synthesis has been reported to be sequentially regulated. Evidence for a coordinate regulation of viral polypeptide synthesis has emerged from the analysis of the rate of protein synthesis taken several times after infection and from the study of viral polypeptide synthesis in the presence of inhibitory concentrations of cycloheximide or puromycin (242,361). The viral polypeptides could be separated in three groups designated α, β and γ. The synthesis of α polypep-

TABLE VI. Molecular Weight of Virus-Infected Cell Polypeptides
Induced by Herpes Simplex Virus (strains F_1 and 17 mp),
Cytomegalovirus (CMV), and Equine Herpes Virus (EHV_1)

HSV_1 (F_1)	HSV_1 (17 mp)	CMV	EHV_1
275	273		>220
260			
	248		
	239	235	
			220
		205	210
			200
			195
			185
177	175		170
157	155	150	148
149˙	145	142	146
130	136	135	
129	129	130	130
126	122		
119	117		115
115	110	105	112
100	100	98	110
94			
91		91	92
88	87/85		
86		85	
82	82/81		80
78		78	
76		75	
73	74	73	72
70		70	
69			69
68	67	68	
	65/64	65	65
62			63
59		58	
57	57	57	57
55			56

TABLE VI. (continued)

HSV$_1$ (F$_1$)	HSV$_1$ (17 mp)	CMV	EHV$_1$
51	51	50.5	52
			49
		48	48
47			46
	43	45	44
39		40	40
36	37	37	
		34	33
		32	
		31	
	28	28	
25		23.5	
	21	21	21
		19.5	
			17.5
	16.5		17.0
	15	15	
	14	13.5	14
	13		
	12.5		
	11		

tides does not require prior protein synthesis in the infected cells and appears immediately after withdrawal of the drug, leading to the conclusion that the corresponding a (IE) mRNA has been accumulated in the presence of the drug (see Section III). In the corresponding untreated cells, their rate of synthesis is maximal at 3 and 4 hours postinfection and decreases thereafter (Fig. 33). The synthesis of β polypeptides requires both the synthesis of a polypeptides and transcription of viral DNA (361). If removal of cycloheximide is concomitant with addition of actinomycin D, no synthesis of β polyeptides occurs. In untreated infected cells, the maximum rate of synthesis of these viral polypeptides takes place between 5 and 7 hours postinfection, at a time where a polypeptide synthesis is greatly reduced

TABLE VII. Molecular Weight (MW) of Virus-Infected Cell Polypeptides Induced by Herpes Simplex Virus

ICP(1)	MW[a] (daltons × 10⁻³)	ICP(2)	MW[b] (daltons × 10⁻³)	ICP(1)	MW[a] (daltons × 10⁻³)	ICP(2)	MW[b] (daltons × 10⁻³)
1	221			20	77	26	76.5
2	205–196	1–4	>200	21	72.5	27	74
3	194			23	71	28	71.5
4	170,165,163	5–8	169,177			29	70
5	151,149	9	156,153	24	68,67.5	30	68
6	146,143	10	145,147			31	66
7	139	11	136	25	64,63		
7.5	132,130	12	132	26	61.5,60	32	61
8	128	13	127	27	58,56.5		
9	122,119	14	120	28	56		
10	117	15	119	29	55,54.5	33	55.5
11	114			31	52	34	52
12	111			32	51.5		
13	109	16	109	33	50		
14	106	17	107	34	49.5	35	49.5
15	103	18	96	35	46.5,45	37	45.3
16	101	19	95	36	42.5	38	42
17	92	20	92	37	39	39	38.5
		21	90	38	37	40	37.5
18	88,85	22	89	39	36,35	41	37
		23	85	40	33.7	42	34.5
		24	83	41	32	43	33
19	78	25	80			44	28
						45	28
				43	26.5	46	26.6
				44	24.5	47	24.5

[a] Data from Bookout and Levy (362).
[b] Data from Morse et al. (367).

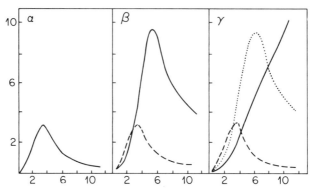

FIGURE 33.   Synthesis of infected cell polypeptides. The amount of three polypeptides representative of (ICP$_4$), a (ICP$_6$), and γ (ICP$_5$) groups of viral-induced polypeptides has been estimated at different times after infection. The results are expressed as the percentage of total amino acid incorporation in each polypeptide at different times after infection. Data from Plummer et al. (74).

(see Fig. 33). The third class (γ) of viral-induced polypeptides is synthesized at an increasing rate until 12 hours postinfection, requires both transcription of viral DNA and prior synthesis of β polypeptides, and has been found to decline at a late time. Therefore, a cascade mechanism for HSV protein synthesis has been described in which synthesis of one class of polypeptides (a) is preceding and regulating the expression of a second group (β), which is itself involved in the expression of the third class (γ) of viral polypeptides (242).

   Inhibition of viral DNA synthesis by cytosine arabinoside was reported to interfere with the transition process from early to late protein synthesis (187). The existence of a true late class of viral polypeptides whose synthesis is not detected in the absence of viral DNA synthesis has been reported by Powell et al. (244). A possible role for viral DNA replication in the regulation of transcription has been suggested in several cases (242–245) while the different patterns of mRNA hybridization reported in

Section III for early and late transcription suggest that the transition process is not the simple result of an increased number of viral genomes transcribed in the infected cells. The cascade regulation process suggests, therefore, that a polypeptides are acting at the level of transcription. Consistent with this hypothesis is the observation that a polypeptides are translocated to the nucleus after they have been synthesized (181,362).

When the infected cells are grown in the presence of the inhibitor canavanine (an analog of arginine), only the a polypeptides and one or two β and γ polypeptides are synthesized (181,363). Viral mRNA made under the same conditions was found to contain transcripts that arise from late regions in untreated infected cells (196). These observations led to the suggestion that more than one polypeptide in each of the a and β polypeptide groups is involved in the transcription of viral DNA, which was reported to precede the transitions from a to β and β to γ polypeptide synthesis. Canavanine would allow partial transitions by selectively inhibiting subsets of these polypeptides (196).

Further evidence demonstrating the role of a (or IE) polypeptides in the synthesis of the mRNA species coding for β polypeptides has been obtained by a study of the induction of the a (IE) mRNA species coding for the enzyme pryimidine deoxyribonucleoside kinase (dPyk) (212,232) and by a study of the pattern of polypeptide synthesis in DNA-negative thermosensitive mutants of HSV_1 (213,364). The results obtained showed that an a polypeptide of molecular weight 175,000 daltons is directly involved at a transcriptional level in the switch from a to β polypeptide synthesis. In tsK (Glasgow strain) and tsB_2 (Houston strain) infected cells, grown at the nonpermissive temperature, the control process is not effective, leading to a restricted pattern of mRNA and protein synthesis similar to that observed in the infected cells grown in the presence of cycloheximide. Interestingly, both impaired migration to the nucleus and aberrant processing of molecular weight 175,000 polypeptide, which were observed in tsK and tsB$_2$–infected cells grown at high temperature, were also observed when wild-type infected cells were grown in the presence of canavanine (181). This provided direct evidence for a viral protein involved in a transcriptional control mechanism in eucaryotic infected cells.

It is interesting that the vast majority of the HSV-specified a early proteins are DNA binding proteins, and therefore, might serve in functions involving interactions with nucleic acids. It has been shown that more than 30% of all the DNA binding proteins isolated from infected cells are a polypeptides. On the other hand, only 10 to 20% of the γ polypeptides have affinity for DNA (362). A selective accumulation of some viral proteins within the cytoplasm or the nucleus of the HSV-infected cells has also been reported. The DNA binding proteins were found to accumulate in the nucleus, reinforcing the hypothesis that they may be involved in some aspect of nucleic acid metabolism (362). The apparent molecular weight of a polypeptides has been found to range between 175,000 daltons and 12,000 daltons, including polypeptides with molecular weight of 136,000, 110,000, 87,000, 68,000, and 63,000 daltons.

B.   LOCALIZATION OF DNA SEQUENCES SPECIFYING VIRAL POLYPEPTIDES ON THE HSV GENOME

An analysis of intertypic recombinant (HSV$_1$ × HSV$_2$) genomes has been made possible because the physical maps of HSV$_2$ and HSV$_1$ DNA are collinear and exhibit different endonuclease restriction sites (see Section II). The comparison of DNA sequence arrangement and the polypeptide pattern induced by intertypic recombinants made it possible to map on the genome the DNA sequences coding for several polypeptides (see Fig. 34). This strategy has been successfully used in the analysis of adenovirus-induced polypeptides (365,366). As this method relies on the analysis of different polypeptide migrations in po-

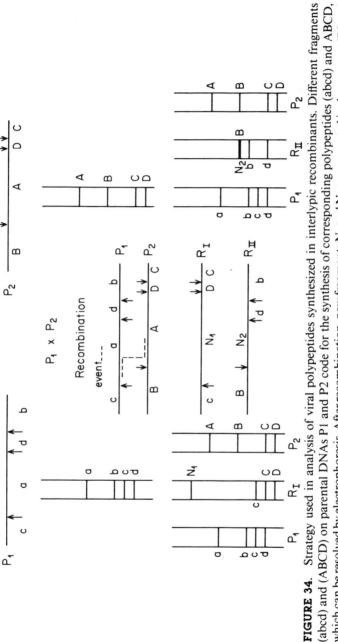

FIGURE 34. Strategy used in analysis of viral polypeptides synthesized in interlypic recombinants. Different fragments (abcd) and (ABCD) on parental DNAs P1 and P2 code for the synthesis of corresponding polypeptides (abcd) and ABCD, which can be resolved by electrophoresis. After recombination, new fragments N₁ and N₂ are generated in the progeny (RI and RII). Comparison of the polypeptides induced by the recombinants and the parents will allow the determination of the site of recombination on the parental DNA molecules.

133

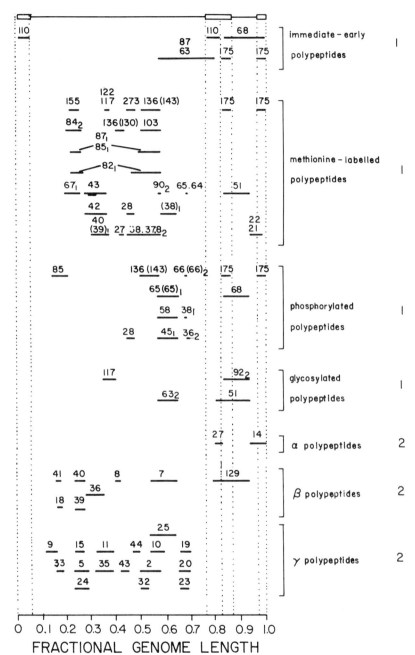

FIGURE 35. Composite map for the location of HSV-induced polypeptides. The location of the different classes of polypeptides has been determined by analysis of intertypic HSV₁ × HSV₂ recombinants. Data (1) from Marsden et al. (359) and (2) from Morse et al. (367).

lyacrylamide gel electrophoresis, it does not strictly map the corresponding structural genes but, rather, the DNA sequences involved in the mobility properties of the considered polypeptides. This point must be borne in mind, expecially in the case of HSV polypeptides that are submitted to several post-translational processing events such as phosphorylation, glycosylation, or sulfation (see above).

The results obtained independently by several laboratories (359–368), allow physical map coordinates to be assigned to several polypeptides of a, β or γ classes (Fig. 35). The salient features of these maps are as follows:

(a) Polypeptides are encoded by genes from all regions of HSV DNA (long and short unique sequences and repetitive sequences).

(b) a Polypeptides are coded by sequences of both long and short regions of the genome. Some polypeptides such as ICP_4 (Vmw 175,000) are coded by repetitive sequences bounding the short and long region. The segregation patterns obtained for ICP_4 show that the gene coding for this polypeptide is located on the map positions 0.93 to 1.00. ICP_{27} would map within the right end of the L sequences, between 0.787 and 0.83 map units. It is of interest that these DNA sequences were shown to specify a(IE) RNA (see Section III).

(c) The templates specifying β and γ polypeptides are randomly distributed throughout the long unique region (U_L) of the genome. The two regions that are essentially involved (located between 20 and 30% and between 50 and 60% from the left hand of the P arrangement) were reported to code for large mRNA species (245).

(d) No major structural protein has been mapped in the short region, whereas several of them map within the long region.

(e)   Glycoproteins and phosphoproteins do not seem to be coded by slustered sequences on the genome.

This procedure allowed also the mapping on the genome of some polypeptides involved in particular functions. It has been possible, therefore, to map the locus responsible for HSV DNA polymerase and viral-induced thymidine kinase (see below), and the functions involved in the increased host synthesis shut-off, induced by HSV_2. These functions lie between 0.52 and 0.59 map units on the genome. The random distribution of β and γ polypeptide templates has led Morse et al. (367) to suggest that β and γ templates are segregated on opposite strands of the DNA. The observations that trace amounts of a RNA hybridize to sequences corresponding to β and γ templates (119) and that large amounts of RNA have been found to be self-complementary (217,369) would support this hypothesis. The finding that sequences involved in the synthesis of polypeptides such as ICP_4 (molecular weight 175,000) lie in the repetitive regions binding the short unique region implies genetic consequences that will be discussed in the following section.

V.   GENETIC ANALYSIS OF HERPES SIMPLEX GENOME ORGANIZATION

A.   GENERAL CONSIDERATIONS

Genetic analysis has proved to be a very powerful tool to progress in the knowledge of the procaryotic genome organization. It is now widely used to define the viral functions that govern the interactions of animal DNA tumor viruses and their respective hosts. The isolation of mutant viruses, unable to complete their lytic cycle or to interact with the cell biology, allows definition of the functions involved in these processes. Therefore, isolation of virus mutants altered in one of the essential steps in the lytic cycle (i.e., penetration, DNA replication, DNA transcription, or translation of viral messages) or in the functions required for efficient inhibition of host synthesis, cell fusion, release of viral particles, etc. is conceivable.

The different mutants of HSV_1 and HSV_2 that have been isolated include mostly ts mutants, altered in many various functions such as DNA replication, control of transcription, synthesis of envelope, or encapsidation process. These mutants have been mapped by genetic methods into recombination and complementation groups and located on the HSV genome by physical mapping.

B. ISOLATION AND CHARACTERIZATION OF HSV MUTANTS

The lack of adequate, simple methods of selection for the isolation of herpes simplex virus mutants unable to achieve their replicative cycle has led the different laboratories involved in genetic studies to search for temperature sensitive mutants. Such mutants, which have been widely used to define essential functions of procaryotes, are usually identical to the wild-type strain at the permissive temperature (31° to 33°C), whereas they are not able to give infectious progeny at the restrictive or nonpermissive temperature (39° to 40°C). Several independently isolated ts mutants have been isolated from different wild-type HSV strains after mutagenesis with 5′ bromodeoxyuridine (Budr) (370–379), nitrous acid (380), 2-aminopurine (AP) (cited in 381), or nitrosoguanidine (NG) (373,378). Before an examination of the behavior of these mutants, several considerations must be pointed out. First, most of the mutants available have been obtained after mutagenesis with Budr, which may be incorporated instead of thymidine in newly synthesized DNA. Therefore, such regions of the DNA represent preferential targets for such a mutagen, leading to possible "hot mutational regions." The second point that we should bear in mind along with the analysis of the ts mutants concerns the nature of the defect or defects that are necessary to induce a thermolabile function. It is generally assumed that ts mutations are due to altered polypeptides whose conformation at high temperature is such that these polypeptides are unable to play their physiological role (either enzymatic or structrual), although the isolation of a ts mutant with a defect in tyrosine tRNA (382) has shown that this is not a rule. The selection of ts mutants may also give rise to clustered mutations if the different domains of a given polypeptide do not play an essential function in the maintenance of the native conformation. It is therefore possible that "hot regions," particularly those involved in the thermal stability of the functional polypeptide, will be affected in ts mutants. Finally, although ts mutations may theoretically occur in all the HSV genes, only those that code for essential functions will be identified; ts mutations may not be isolated for functions inessential for virus multiplication, under the laboratory conditions used. This is the case for thymidine kinase activity (383,384) and glycoprotein gC (385,387). Therefore, the biological significance of the frequency at which a function appears to be altered among the ts mutants studied is subject to caution. For example, a brief overview of the ts mutants isolated by several laboratories reveals that about 50% of the mutants are DNA(-), and the conclusion drawn from such observations is that most of these mutants have been isolated under very similar conditions. It thus becomes necessary to isolate different lethal conditional mutants (i.e., cold-sensitive mutants) and to develop the use of different mutagens in the search for ts mutants or to develop site-directed mutagenesis in order to avoid mutational artifacts (381).

Host range mutants, expressing their temperature-sensitive

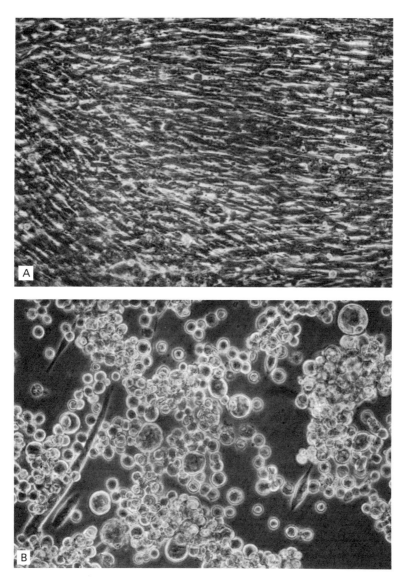

FIGURE 36. Cytopathic effect induced by herpes simplex type 1 virus. *A;* Uninfected BHK (C13) cells; *B;* BHK (C13) cells infected with a HSV1 strain (P431) freshly isolated from a patient. This picture was taken after 18 hours' incubation at 37° C.

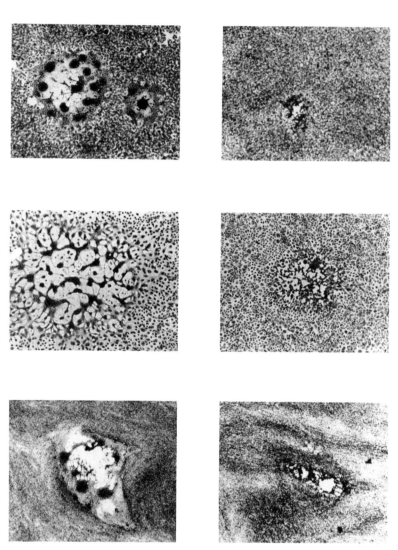

FIGURE 37. Cytopathic effect induced by strain 17 syn⁺ and strain 17 syn⁻ with different cell lines. *Top,* Vero cells; left, HSV₁ syn⁻; right, HSV₁ syn⁺. *Middle,* CV1 cells; left, HSV₁ syn⁻; right, HSV₁ syn⁺. *Bottom,* BHK (C13) cells; left, HSV₁ syn⁻; right, HSV₁ syn⁺. Photographs provided by A. Taylor.

140

phenotype only in certain types of cells, have also been described (378,388,389).

Among the other types of mutants isolated thus far are those that modify what has been called the "social behavior of the infected cells" (390). Upon infection with HSV, most of the cells become rounded and clump (see Fig. 36). However, several variant and mutant strains have been reported to induce fusion of the infected cells, leading to the formation of polykaryotes (Fig. 37) (372,390–396). The genetic locus involved in cell fusion of the infected cells has been called "syn"; the syn+ form is that which induces rounding and clumping. Most of the "wild type" strains isolated on patients exhibit a syn+ phenotype.

Mutants of viral enzymes have also been described. Among them are mutants selected as being resistant to the drug phosphonoacetic acid (PAA) (243,312,313). Such mutants were shown to induce altered DNA polymerase activities in the infected cells (312). Interestingly, a thermosensitive mutant of HSV$_1$ called ts D9 (368), which was not selected for PAA resistance, was found to be resistant to this compound (397). Purification of the DNA polymerase induced by this ts mutant and by a ts+ revertant (called DR3) of this mutant demonstrated, that the DNA polymerase encoded by ts D9 is thermolabile and PAA-sensitive (398). These results are in agreement with the observation that the DNA polymerase activity induced at nonpermissive temperature encoded by ts D9 is thermolabile and PAA-sensitive (398). These results are in agreement with the observation that the DNA polymerase activity induced at nonpermissive temperature by ts D9 in vivo is thermolabile (399). These results indicate that the genetic locus altered in PAAr mutants is the structural gene for viral DNA polymerase and therefore permits location on the HSV genome of this essential function for viral multiplication. Mutants of the viral enzyme thymidine kinase (TK) have also been obtained. Interestingly, a chain termination mutant has been obtained by selection of a virus resistant to Budr when grown in thymidine-kinase-deficient (LMTK⁻) cells (379). Such selective procedure could be applied successfully because, as already mentioned, TK activity is not required for virus multiplication under most laboratory conditions. TK⁻ mutants are also able to give plaques in the presence of thymidine arabinoside (araT), an inhibitor of herpes simplex replication (400).

The chain termination mutant isolated by Summers et al. (379) has been reported to be suppressible *in vitro* by yeast suppressor tRNAs (W.P. Summers and W.C. Summers, abstract of the 9th Annual ICN-UCLA Symposium. Molecular and Cellular Biology, Keystone, Colorado, 1980. Revertants were selected in LMTK⁻ cell grown in the presence of amethopterin to inhibit *de novo* thymidylate synthesis, hypoxanthine as a source of purines, and thymidine (HAT medium). These revertants have regained a TK⁺ activity and have induced the synthesis of a polypeptide having a molecular weight of about 40,000 daltons, which was missing in the chain termination mutant (379). The molecular weight of TK polypeptide has been reported to be 43,000 to 44,000 daltons (345,401) by polyacrylamide gel electrophoresis. The use of such potential nonsense mutants may allow the isolation of eucaryotic cells carrying nonsense suppressors similar to those described in procaryotes. Coen and Schaffer (abstract of the 9th Annual ICN-ULCA Symposium, Keystone, Colorado, 1980) reported the isolation of an HSV₁ mutant resistant to cycloguanosine (ACG) at 39°C, with a thermolabile TK activity. The location of the defect involved is in agreement with the known location of TK gene on the genome. A second locus for ACG resistance was found to be linked to PAA resistance, suggesting that resistance to both drugs may also arise from a single mutation in the viral DNA polymerase coding sequences.

C.   CONSTRUCTION OF THE HSV GENETIC MAPS

Several methods have been used to map on the HSV genome the different loci responsible for the mutant phenotypes described above. They include classical genetic methods such as recombination and complementation analysis and more biochemical approaches such as marker rescue experiments with intratypic or intertypic fragments of DNA.

1. Rationale of Genetic Mapping By Recombination and Complementation Analysis

a. Recombination

☐ *The early studies of phage geneticists, performed in the years 1946–1949, provided the basis for utilization of recombination in the analysis of genome organization (402,403). Two types of crosses have been widely used in genetic analysis: two-factor crosses and three-factor crosses. The basic feature of these tests are outlined below.*

i. TWO FACTOR CROSSES. *Let us consider two distinct loci (*a,b*) located at different sites on a DNA molecule, and let us call* a⁺ *and* b⁺ *the wild-type dominant alleles of the mutants* a_1, a_2, . . ., a_n; b_1, b_2, . . ., b_n *which exhibit the (*a⁻*) and (*b⁻*) phenotypes. Under conditions that lead to recombination of two DNA molecules bearing different mutations in the* a *and* b *loci, the progeny will be as indicated on page 143.*
The frequency with which the recombinants will occur is a function of the probability at which a crossover between the two loci on the DNA may take place. In other words, loci located far apart will give rise to high frequences of recombination, whereas very close loci will rarely segregate. This is the basis for two-factor crosses. In the case of virus crosses, the susceptible cells are infected with two different mutant strains (usually with an equal number of infectious particles from both types), and incubated under conditions that allow the recombination of the parental DNA. Scoring

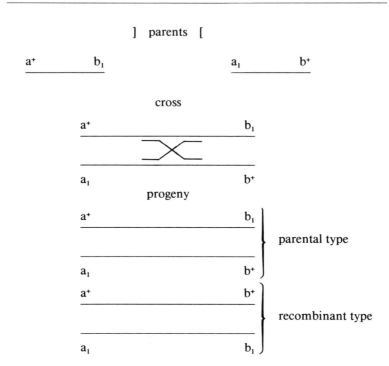

of the progeny will allow the establishment of recombination frequency and, therefore, the estimation of the distance
between pairs of loci. In the case of herpes simplex ts mutants, the mixed infected cells are grown at the permissive
temperature and the frequency of recombination is obtained by testing the ability of the progeny to plate at the
nonpermissive temperature at which it selects for ts^+ recombinants. It should be pointed out that one cannot simply
distinguish among the progeny, the double ts recombinant
(equivalent to a_1b_1) from the single ts mutants. Therefore,
in practice, the assumption is usually made that the dou-
ble-mutant recombinant class occurs at the same frequency
as that of wild-type recombinants (equivalent to a^+b^+) and
the recombinant frequency is taken as twice the frequency

TABLE VIII. Geometric Mean of Three Sets of Recombination Frequencies Obtained with HSV$_1$ (17) ts Mutants[a]

	A	B	C	D	E	F	G	I	J
A	0	0.08	14.49	9.22	2.00	28.70	7.69	0.136	2.28
B		0	2.20	11.86	<0.013	20.94	0.164	0.385	5.19
C			0	12.16	3.53	8.17	5.84	13.03	4.66
D				0	4.63	49.50	8.20	14.86	13.65
E					0	18.03	0.49	5.27	2.88
F						0	49.80	11.10	8.58
G							0	0.38	3.10
I								0	3.10
J									0

[a] Data taken from Brown (404).

145

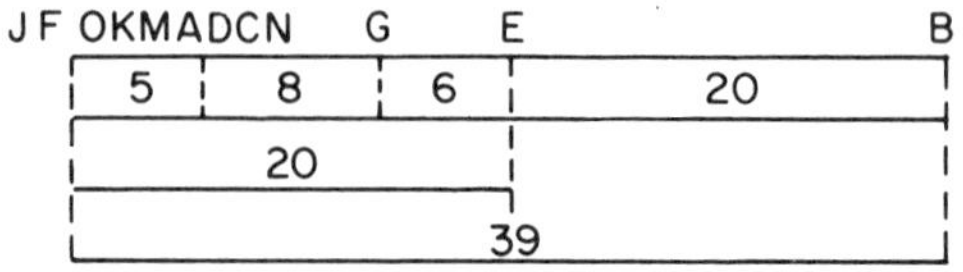

FIGURE 38. Genetic map for HSV_1 (KOS). The recombination frequencies obtained in two-factor crosses are indicated. Data from Parris et al. (423).

of wild-type recombinants (ts^+). When the recombination frequencies obtained with different pairs of mutants are additive, the construction of a genetic map is possible (see Fig. 38). However, in many cases, the genes' order cannot be deduced from the recombination frequencies obtained in a two-factor cross. The introduction of a third marker allows to perform a three-factor cross and to determine the position of two markers relative to the third one.

For instance, the ordering of HSV (strain 17) tsA, tsC, and tsD based only on the recombination frequencies obtained in the crosses A×C, and A×D, and C×D (Table VIII) leads to an ambiguity. Thus, A×D cross gives 9.22% ts^+ and A×C 14.49% ts^+. Two possible locations may be drawn:

(1) A 9.22 D C

14.49

(2) C 14.49 A 9.22 D

We expect that the result from the C×D cross will allow these two possibilities to be distinguished. The frequency of ts⁺ should be about 5% in case 1 and is expected to be 23% in case 2. The result obtained is 12.16% and therefore does not allow the decision whether the order is as (1) or (2).

Making the assumptions that (a) crossover events occur at random on the DNA molecule; (b) crossover will occur more frequently when markers are farther apart; and (c) double crossover occurs much less frequently than single crossover, we can establish the order of two distinct loci with respect to a third nonselected marker.

ii. THREE FACTOR CROSSES. *Let us consider the two possible arrangements* a b c *and* b a c *for three independent markers and examine the possible progeny of a cross involving the following parents P_I (a ⁻b ⁺c ⁻) × P_{II} (a ⁺b ⁻c ⁺).*

In the first case (abc), *the selection of wild-type recombinants* a⁺b⁺ *(selected markers) will lead predominantly to* a⁺b⁺c⁻ *progeny:*

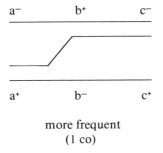

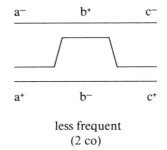

more frequent   less frequent
(1 co)   (2 co)

In the second case (bac), most of the a^+b^+ recombinants will also be c^+:

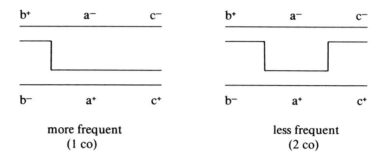

more frequent	less frequent
(1 co)	(2 co)

The morphology marker (syn) has been used as a non selected marker in three-factor crosses with HSV mutants (372,404). For example, the order of tsA, tsD, and tsC loci may be deduced from the results in Table IX. The cross syn tsA×syn$^+$ tsD gives 18 syn for 34 syn$^+$ among the selected ts$^+$ recombinants. The two possible arrangements are:

(1)   syn   tsa   +

         +   +   tsD

(2)   syn   +   tsA

         +   tsD   +

Only arrangement (1) is compatible with the higher frequency of syn$^+$ among ts$^+$ recombinants. Similarly, the cross syn—tsD×syn$^+$ gives 50 syn and 90 syn$^+$ among the ts$^+$ recombinants, and the cross syn—tsA×syn$^+$—tsC gives 30 syn and 160 syn$^+$ among the selected ts$^+$ recombinants.

It is concluded that the markers' order is syn tsA, tsD,

tsC. The map presented in Fig. 39 has been built up in this way.

b. Complementation. *Theoretically, the recombination event may take place at the level of a single base pair and, therefore, should allow the ordering of mutated loci even if they are very close to one another. In practice, several considerations make it impossible to map one mutant relative to the other because the recombination frequencies are too small to be considered with confidence. Also, the recombination test does not allow us to estimate the size and number of the functional units involved in the expression of a given phenotype. On the other hand, the use of complementation tests leads to the classification of different mutants in functional units which have been called cistrons after the early work of Benzer, who first introduced the cis-trans (complementation) test in his study of the rII mutants of bacteriophage T_4.*

Let us consider different functions A and B involved in a common biological pathway (P). These functions are coded by different genes (a,b) on the genome. Several mutants M_1, M_2 and M_3, M_4 are available and all exhibit the same phenotype (P⁻).

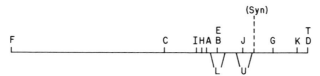

FIGURE 39. Genetic map for HSV₁ (17). This map has been established after three-factor crosses (406). The position of syn marker has been indicated in brackets due to the multiple possible sites for syn mutants to occur (see text). Mutants L and U could not be unequivocally mapped.

TABLE IX. Segregation of the syn and syn⁺ Marker in Three-Factor Crosses Performed with HSV_1 (17) Mutamts[a]

syn \ syn⁺	A 0/0/0	B 0/0/0	C 0/30/0	D 0/0/0	E 0/0/0	F 0/30/0	G 0/0/0	I 0/0/0	J
A 0/0/0		20/10/2	30/160/10	18/34/2	22/7/1	15/135/3	4/10/0	27/4/0	
C 200/0/0	150/48/5	380/28/20		23/2/1	280/180/20	11/51/3	190/70/12	231/72/5	
D 12/0/0		95/55/10	50/90/11		23/9/0	20/90/9	98/20/4	170/30/10	
E 0/0/0	5/30/0	0/0/0	11/47/2	8/47/3		20/80/2	6/28/2	10/40/3	
F 100/0/0	40/16/3	50/14/2	11/0/0	115/25/5	38/20/4		100/20/10	160/40/8	
G 0/0/0	48/3/2	50/14/2	140/200/20	12/28/5	60/20/5	20/60/9		80/35/5	
I 0/0/0	5/30/4	220/23/10	20/60/1	25/70/5	40/5/5	7/28/2	30/90/5		
J 0/0/0	4/21/4	30/50/10	30/140/10	20/80/6	10/22/5	8/58/8	10/60/5	50/120/11	

[a] The segregation of syn and syn^+ pheontype has been scored among ts^+ recombinants after a cross between ts mutants. The results $n_1/n_2/n_3$ represent syn, syn^+, and mixed syn/syn plaques. Data taken from Brown (404).

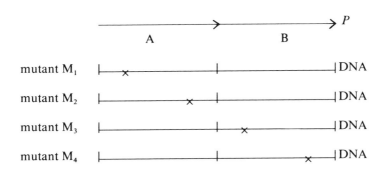

Mutants M_1 and M_2 which are defective only in function A and express unaltered function B are able to supply the defective B function of mutants M_3 and M_4. On the contrary, mutants M_1 would be unable to "complement" mutant M_2 because each of them is altered in the same functional unit. The same conclusion is drawn for M_3 and M_4.

In other words, mixed infection of a cell under restrictive conditions (to avoid virus multiplication) with two mutant viruses bearing mutations in different functional units may result in the production of progeny, both mutants supplying the function that is defective in the other.

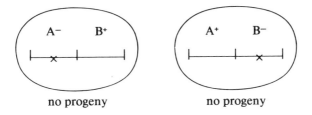

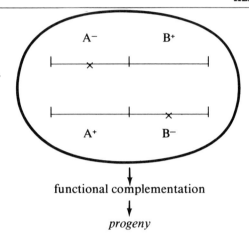

functional complementation

progeny

The complementation test is valid only if recombination does not take place and if the mutations studied are recessive over the wild-type allele. Therefore, all experiments should include a control test in which cells are mixedly infected with a wild-type virus and a mutant virus bearing both mutations on the same DNA molecule (the mutations being arranged in cis*) to control that under conditions where the mutant functions are complemented by the wild-type functions, the virus is able to replicate and generate progeny. The* trans *test then allows us to assign mutants to different complementation groups which correspond to distinct functional units (also called cistron).*

test *cis–trans*

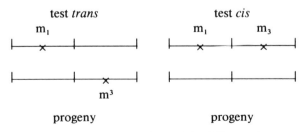

mutations m_1 and m_3 are located in different cistrons

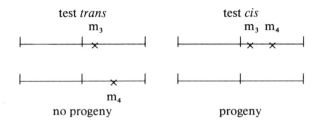

mutations m_3 and m_4 are located in the same cistron

In practice, the situation is slightly complicated by the possibility that recombination takes place (see text) and that intracistronic complementation occurs, especially in the case of multimeric proteins made by the assembly of several polypeptidic chains.

Several complementation tests have been used to screen the different HSV_1 and HSV_2 mutants isolated thus far.

i. QUALITATIVE COMPLEMENTATION TEST (381). *Filter paper disks are saturated with mixtures of the mutants employed and placed on monolayers of monkey vero cells. A monolayer of methyl cellulose (2%) is poured on the cells which are incubated for 5 days at the nonpermissive temperature. When complementation occurs, clear areas appear on the monolayer after staining with neutral red. The complementation index (CI) is calculated as*

$$CI = \frac{(A + B)\, npT/pT}{1/2\, (AnpT/ + BnpT/pT)}$$

where A and B represent the ts+ obtained with A and B mutants.

npT = Nonpermissive temperature; pT =permissive temperature. □

ii. INFECTIOUS CENTER ASSAY (372). *A mixture of mutant virus is used to infect baby hamster kidney cells (BHK-21 C13). After neutralization of the unadsorbed virus by the addition of human serum (10%), dilutions of the infected* *cells are mixed with 4×10^6 uninfected cells and grown at both restrictive and permissive temperatures. When complementation occurs, the infected cells produce an infectious center plaque at the nonpermissive temperature.*

iii. QUANTITATIVE (YIELD OF PROGENY) COMPLEMENTATION TEST (372,381). *In this test, the yield of progeny, obtained in mixed infections, is compared to that obtained with single mutant infections.*

Monolayers or tube cultures are infected with 2.5 or 5 pfu of each ts mutant per cell and incubated for 18–20 hours at the nonpermissive temperature. The progeny virus is released by isolation of three cycles of freezing and thawing and titrated at the permissive temperature.

In all three cases, control experiments using infections with single ts mutants are performed under similar conditions.

The complementation indice used by Schaffer et al. (381) is

$$CI = \frac{(A + B) \, npT}{(A \, npT + B \, npT)}$$

whereas Brown et al. (367) use the following formula:

$$CI = \frac{(A+B) \, pT - (A+B) \, npT}{1/2 \, (A \, pT + B \, pT)}$$

in both cases a CI of 2 or greater is taken to indicate positive complementation. The results obtained are reported in Tables X and XI.   □

2. RECOMBINATION AND COMPLEMENTATION ANALYSIS OF HERPES SIMPLEX GENOME

a. **Recombination Analysis.** Two- and three-factor crosses have been used to construct genetic maps for different strains of HSV_1, (372, 404–408). The map obtained with two-factor crosses for HSV_1, KOS strain, is drawn in Fig. 38; a typical

TABLE X. Comparison of the Complementation Indices (CI) Obtained by Infectious Center Assay and Progeny Yield Test with HSV_1 (17) ts Mutants[a]

	A	B	C	D	E	F	G	I	J
A	1 / *1*	2.1 / *4.7*	2.0 / *3.2*	9.8 / *22.8*	8.5 / *1.1*	3.6 / *2.0*	9.0 / *2.7*	36.4 / *14.5*	0.7 / *3.7*
B		1 / *1*	2.6 / *2.5*	6.7 / *2.8*	0.7 / *1.1*	19.0 / *19.8*	34.7 / *14.6*	161.9 / *18.7*	161.2 / *85.2*
C			1 / *1*	3.8 / *7.0*	2.0 / *3.7*	12.9 / *10.4*	4.7 / *4.9*	5.8 / *9.3*	3.3 / *5.0*
D				1 / *1*	30.9 / *34.5*	4.5 / *9.5*	27.1 / *13.9*	11.0 / *9.6*	7.0 / *17.2*
E					1 / *1*	3.1 / *50.5*	38.1 / *4.1*	51.9 / *36.1*	55.6 / *11.9*
F						1 / *1*	3.2 / *10.1*	2.1 / *11.5*	3.3 / *14.1*
G							1 / *1*	2.3 / *3.9*	1.6 / *2.9*
I								1 / *1*	18.7 / *14.1*
J									1 / *1*

[a]Numbers in italics refer to progeny yield test. Data taken from Brown (404).

155.

TABLE XI.   Complementation Analysis of HSV₁ (17) Mutants[a]

	H	K	L	S	T	U
H	1	57 *11,68*	15,18 *46,84*	162 *38*	280 *263*	508 *80*
K		1	13,33 *3,61*	106 *188*	6.4 *2.3*	28.5 *110*
L			1	3.2 *13.8*	14 *45*	0.34 *0.10*
S				1	67 *1388*	31.5 *98*
T					1	28.4 *70.2*
U						1

[a]The infectious center assay method (cf. text) has been used. The results are expressed as the C.I. obtained in two independent determinations. Self crosses have been performed in the same way. Data taken from Crombie (406).

series of recombination frequencies obtained for the Glasgow strain 17 is reported in Table XII. The plaque morphology marker (syn) has ben used to perform three-factor crosses and to construct the map of HSV₁ (17 mp) presented in Fig. 39. As already pointed out by Ritchie (409), the construction of these maps relies on recombination frequencies that are subject to various sources of variability and a standard cross in each recombination experiment needs to be included.

The unique structural characteristics of the HSV genome (see Section II) allow us to predict the following.

(a)   In each of the two unique regions (U_L and U_S), the genes will map linearly to one another.

TABLE XII.   HSV-$_1$ Complementation Groups[a]

| | Different ts Mutants Isolated From Strains | | | | | | Comple-menta tion Groups |
KOS	KOS 1.1	KOS pp 601	17	13	HFEM (1)	HFEM (2)	
A1	656				N103		1-1
B2	901		D			LB2	1-2
C4	833,661	3					1-3
D9	84	3					1-4
E6						LS2	1-5
F18						LS1,LS2	1-6
G3				C4		LB1	1-7
I11							1-8
J12	822		I	G5	B5		1-9
K13							1-10
L14			F			LB3	1-11
M19							1-12
N20			G				1-13
022	661	3					1-14
P23							1-15
			A				1-16
	478						1-17
		7					1-18
					B1		1-19
					B7		1-20
						LB4	1-21
						LB5	1-22
				D10			1-23

[a] Data taken from Schaffer et al. (381).

(b)   All the loci in U_L will map equidistant from all those in the U_s segment.

(c)   If the repetitive regions correspond to a recombinational hot zone, all the genes located in the U_L region will appear loosely linked (independent) to those located in U_s region.

(d)   The different size of U_L and U_S portions of the genome will statistically favor the isolation of mutants in U_L (U_L is about seven times longer than U_S).

b. **Complementation Analysis.** Several complementation maps have been established in different laboratories which have led to the identification of 10 cistrons for HSV_1 strain 17 (406), 15 cistrons for HSV_1 strain KOS (373), and 18 cistrons for HSV_2 (410). A collaborative study performed by Schaffer et al.

TABLE XIII.   HSV-$_2$ Complementation Groups[a]

Different ts Mutants Isolated from Strains					Comple- mentation Groups
KOS	HSG52	IPB2	333	UW268	
	1				2-1
H9		1,42082		19	2-2
B5	6				2-3
	9				2-4
C2	10			1	2-5
	11		74	5	2-6
A8		42082			2-7
	2				2-8
	8		69,74		2-9
	3				2-10
	4				2-11
	5		69		2-12
	12				2-13
	13	42082			2-14
D6				5	2-15
E7					2-16
F3				33	2-17
G4					2-18
				11	2-19
				12	2-20

[a] Data taken from Schaffer et al. (381).

TABLE XIV. Variability in the Recombination Frequencies Obtained along a Recombination Analysis of HSV_1 (17) Mutants [a]

	A	B	D	F	G	I	J
T	29.4	*20.6*	0	27.6	*17.3*	nd	nd
	9.5	13	nd	9.6	nd	*22.1*	nd
	5.8	30.5	0	29.3	19	12	nd
	12.7	nd	0	nd	*11.8*	*10.4*	nd
	27.6	nd	0	nd	18.5	*26.8*	27.7
L	6	*0.05*	nd	*6*	nd	7	nd
	12.9	*0.13*	nd	20.6	*1.0*	nd	*1.0*
	nd	*0.12*	nd	2.1	nd	nd	nd
U	17.1	*3.5*	nd	28.8	*10*	nd	nd
	24.3	1.9	nd	nd	*4.8*	8	*1.2*
H	11.1	*4.1*	11.4	4.7	12.3	8.1	3.8
	12.1	*1.6*	nd	3.2	14.9	3.2	nd
	4.9	11.9	nd	15.6	14.0	11.4	*6.4*
	nd	6.8	14.3	4.3	nd	nd	nd
	nd	4.6	*11.3*	1.7	nd	nd	nd
K	5.3	10.6	*2.0*	30	10.1	24.9	nd
	5.3	20	*0.63*	0.6	13	1.4	5.8
	8.0	8.8	3.6	2.2	nd	nd	nd
	nd	4.4	2.5	18.5	nd	nd	nd

[a] The recombination frequencies obtained in the two factor crosses have been calculated as follows:

$$R.F. = \frac{100 \times 2 \ (A+B) \ npt/pt}{1/2 \ (Anpt/pt + Bnpt/pt)}$$

where npt and pt represent the nonpermissive and permissive temperatures. The results on one horizontal line come from experiments performed the same day. Several independent results are reported and show the variability of the method. Numbers in italics refer to experiments in which the ratio syn/syn+ was not in the expected 0.3 to 3.0 ratio; nd stands for "no data." Data are from Crombie (406).

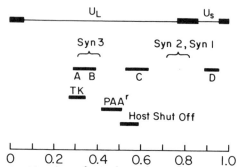

FIGURE 40. Physical map of some HSV functions. The figure summarizes the results obtained by intertypic recombination. Scale shows the fractional genome length.

(381) permitted classification of 43 different thermosensitive mutants of HSV$_1$ and 29 ts mutants of HSV$_2$ in 23 and 20 complementation groups, respectively (see Tables XII and XIII on pages 157 and 158). The most important conclusion from this study is that the majority of the *ts* mutants independently isolated in different laboratories were located in the same functional units (or genes), indicating therefore that the number of cistrons identified to date is very small as compared to the theoretical coding capacity of the HSV genome. This situation may be due to the nature of the mutagen and to the selection procedure used throughout the isolation of mutants. However, such a collaborative study allows us to identify the defect of unknown mutants at the molecular level, when they belong to certain groups. For example, ts mutants of group 1–4 code for a temperature-sensitive DNA polymerase and ts mutants which belong to complementation group 1–9 fail to synthesize the correct amount of gB glycoprotein at the nonpermissive temperature.

Analysis of the progeny obtained in complementation tests with HSV$_1$ reveals that recombinant viruses are produced under the conditions used (373) and that virus particles with an intermediate ts phenotype may be generated in these tests (411).

Nesser and Timbury (412) investigated both intertypic and intratypic complementation tests and confirmed that a considerable proportion of recombinants are produced in both the infectious center and the yield of virus assays. Furthermore, the results obtained by varying the multiplicity of infection suggested that noninfectious particles probably contribute to complementation. The molecular interactions which are involved in HSV complementation would appear, therefore, to be more complex than expected. See Table XIV on page 159.

D. PHYSICAL MAPPING OF HSV MUTANTS ON THE GENOME

The location of many mutants and viral functions on the HSV genomes has been made possible by the use of intratypic (or homotypic) and intertypic marker rescue techniques.

The marker rescue technique has been widely used and described in several different viral systems (413–418). In these experiments, native DNA molecules and selected individual fragments are denatured and annealed in such a way that heteroduplexes are generated *in vitro* and subsequently are used to infect cells accordingly to the calcium phosphate transfection technique of Graham and Van Der Eb (419). The rescue of selected HSV_1 ts mutations by homotypic sheared DNA was first reported by Wilkie et al. (420). Unlike the partial heteroduplex repair technique described above, the rescue of ts mutations by transfection with DNA mixtures of intact HSV DNA and shared DNA (or endonuclease fragments) depends on recombination events, in the transfected cells, between an intact genome and a fragment of DNA bearing the considered allele. This technique has been successfully used to map fine temperature-sensitive mutations on the HSV_1 (strain 17) genome (421) and to confirm the already known location of thymidine kinase structural locus on the genome (422). The results obtained by

Stow et al. (421) did not allow correlation of recombination mapping with physical mapping, whereas a similar physical mapping of HSV$_1$ ts mutants by marker rescue led to a good agreement between the order of the markers obtained by the two techniques in the case of strain KOS (423).

Transfection by DNA mixtures (intact genomes and individual endonuclease restriction fragments) and by partial heteroduplexes have also been used to transfer the PAA resistance marker from one viral genome to another and have allowed performance of a fine mapping of the genetic locus involved in the expression of this phenotype. The *PAAr* locus was mapped between 0.46 and 0.49 map units (424), confirming the results obtained by intertypic marker rescue (425) and reported in the next paragraph.

The use of intertypic recombinants in the analysis of HSV-induced polypeptides has been detailed above. A similar rationale may be used to perform restriction endonuclease analysis of intertypic recombinants produced between HSV$_1$ and HSV$_2$ (intertypic marker rescue). This kind of analysis allowed location, for example, of the tsD and tsK mutations of HSV$_1$ (strain 17) within both the repeats flanking the short unique region (U$_S$) of the DNA, whereas tsS, tsA, and tsH were found to map within the long unique region (U$_L$) (426). Intertypic marker rescue has also been used to locate several functions such as virus-induced cell fusion, viral-induced shutoff of host protein synthesis, viral thymidine kinase (367), and PAA resistance (367,427). These results are summarized in Fig. 40 on page 160.

Interestingly, PAA resistance locus of strain 17 was found to be located between 0.39 and 0.43 map units (427), far apart (about 4 kilobases) from the location reported for the same locus in strain KOS (424,425). If PAA resistance is likely to represent modification of the viral DNA polymerase, these different locations would imply that two distinct loci are involved in DNA polymerase activity (427).

Similarly, the syn phenotype has been found to depend upon

the expression of three distinct loci on the genome (428). Syn 1 and syn 2 loci do not map at a position corresponding to any of the A,B,C, or D major glycoprotein, whereas the syn 3 locus has been located in the region coding for A and B glycoproteins (Fig. 40). The evidence that syn phenotype represents the expression of more than one functional unit may be put forward to explain the discrepancy observed between the physical and the genetic mapping of HSV_1 (strain 17) ts mutants if different syn loci were altered in the ts mutants used in the three-factor cross experiments.

Some other interesting conclusions and developments emerge from the physical mapping performed thus far:

(a) The physical map location of several loci on the HSV genome (strain KOS) allows to determine that markers located in the long unique region (U_L) do not map equidistant to markers in U_S, as the HSV DNA structure led us to predict.

(b) Recombination frequencies are consistent with the physical map location for mutations located in the long unique region (U_L), whereas the recombination frequency for two mutants located in U_L and U_S were always found to be very high. This observation is consistent with the possibility that the joint region is a high recombinational zone of the DNA molecule.

(c) Physical mapping of the ts mutants from different complementation groups confirms that the genes involved in DNA synthesis are not clustered.

(d) A comparison of the physical intertypic marker rescue mapping with the patterns of polypeptides induced by intertypic recombinants allows to map about 30 structural genes and functions.

(e) Physical mapping of ICP_4 polypeptide in the reiterated ca and a'c' regions implies that the HSV genome is diploid for the gene encoding this polypeptide and that mutants unable to synthesize ICP_4 (tsD and $tsLB_2$ mutants of complementation group 1–2) must bear two mutated copies of the gene coding for this polypeptide, a conclusion consistent with the results obtained in intertypic marker rescue experiments performed with these mutants. The implications of this particular situation for HSV_1 DNA replication are discussed elsewhere (429).

(f) The molecular weight of TK-related polypeptides induced by TK mutants was found to correlate with the physical mapping of these mutants, suggesting a leftward direction of transcription of the TK gene on the prototype arrangement (P) of HSV DNA M.H. Wagner, J.R. Smiley, and W.C. Sumers, abstract of the 9th Annual ICN-UCLA Symposium, Keystone, Colorado, 1980). Hybridization experiments performed with cytoplasmic RNA isolated from HSV_1-infected cells and separated strands of a 2.4 Kb EcoR1 cloned fragment containing the coding sequences for TK suggest that portions of this fragment are transcribed from both strands prior to transport into the cytoplasm (D.L. Hare, L.I. Pizer, and J.R. Sadler, abstract of the 9th Annual ICN-UCLA Symposium, Keystone, Colorado, 1980).

(g) The temperature-sensitive lesions corresponding to four ts mutants (*ts*T, *ts*D, *ts*c&5, and *ts*K) have been mapped within the coding sequences of the immediate early polypeptide Vmw 175 (ICP_4). (429)

REFERENCES

1. Roizman, B. and Spear, P.G. (1973), in *Herpes Viruses in Ultrastructure of Animal Viruses and Bacteriophages: An Atlas.* (A. J. Dalton and F. Hagwenau, Eds.), p. 83, Academic Press, New York.

2. Roizman, B. and Furlong, D. (1974), *The Replication of Herpes Virus in Comprehensive Virology,* Vol. 3 (H. Fraenkel Conrat and R.R. Wagner, Eds.) p. 209, Plenum Press, New York.

3. Epstein, M.A. (1962), *J. Exp. Med.* **115**:1.

4. Zambernard, J. and Vatter, A.E. (1966), *Virology* **28**:318.

5. Chopra, H.C., Shibley, G.A., and Walling, M.J. (1970), *J. Microsc.* **9**:167.

6. Cook, M.L. and Stevens, J.G. (1970), *J. Ultrastruct. Res.* **32**:334.

7. Trung, P.H. and Lardener, F. (1972), *J. Microsc.* **14**:271.

8. Furlong, D., Swift, H., and Roizman, B. (1972), *J. Virol.* **10**:1071.

9. Roizman, B., Spring, S.B., and Schwartz, J. (1969), *Fed. Proc. Fed. Am. Soc. Exp. Biol.* **28**:1890.

10. Wildy, P., Russel, W.C., and Horne, R.W. (1960), *Virology* **12**:204.

11. Kazama, F.Y. and Schornstein, K.L. (1973), *Virology* **52**:478.

12. Nayak, D.P. (1971), *J. Virol.* **8**:579.

13. Strandberg, J.D. and Carmichael, L.E. (1965), *J. Bacteriol.* **90**:1790.

14. Nazerian, K. and Witter, R.L. (1970), *J. Virol.* **5**:388.

15. McCombs, R., Brunschwig, J.P., Mirkovic, R., and Benyesh-Melnick, M. (1971), *Virology* **45**:816.

16. Siegert, R.S. and Falke, D. (1966), *Arch. Gesamte Virusforsch.* **19**:230.

17. Falke, D., Siegert, R., and Vogell, W. (1959), *Arch. Gesamte Virusforsch.* **9**:484.

18. Morgan, C., Ellison, S.A., Rose, H.M., and Moore, D.H. (1954), *J. Exp. Med.* **100**:195.

19. Darlington, R.W. and Moss, L.H. III (1969), *Prog. Med. Virol.* **11**:16.

20. Fong, C.K.Y., Tenser, R.B., Hsiung, G.D., and Gross, P.A. (1973), *Virology* **52**:468.

21. McGraken, R.M. and Clarke, J.K. (1971), *Arch. Gesamte Virusforsch.* **34**:189.

22. Epstein, M.A. (1962), *J. Cell. Biol.* **12**:589.

23. Smith, K.O. (1964), *Proc. Soc. Exp. Biol. Med.* **115**:814.

24. Stein, S., Todd, P., and Mahoney, J. (1970), *Can. J. Microsc.* **16**:851.

25. Rubenstein, D.S., Gravell, M., and Darlington, R. (1972), *Virology* **50**:287.

26. Holmes, I.M. and Watson, D.H. (1963), *Virology* **21**:112.

27. Epstein, M.A., Achong, B.G., and Barr, Y.M. (1964), *Lancet* **1**:702.

28. Morgan, C., Rose, H.M., and Mednis, B. (1968), *J. Virol.* **2**:507.

29. Schwartz, J. and Roizman, B. (1969), *Virology* **38**:42.

30. Strandberg, J.D. and Aurelian, L. (1969), *J. Virol.* **4**:480.

31. Jasty, V. and Chang, P.W. (1972), *J. Ultrastruct. Res.* **38**:433.

32. Morgan, C., Rose, H.M., Holden, M., and Jones, E.P. (1959), *J. Exp. Med.* **110**:643.

33. Shimono, H., Ben-Porat, T., and Kaplan, A.S. (1969), *Virology* **37**:49.

34. Olshevesky, J. and Becker, Y. (1970), *Virology* **40**:948.

35. Abodeely, R.A., Palmer, E., Lawson, L.A., and Randall, C.C. (1971), *Virology* **44**:146.

36. Robinson, D.J. and Watson, D.H. (1971), *J. Gen. Virol.* **10**:163.

37. Spear, P.G. and Roizman, B. (1972), *J. Virol.* **9**:143.

38. Heine, J.W., Honess, R.W., Cassai, E., and Roizman, B. (1974), *J. Virol.* **14**:640.

39. Kemp, M.C., Perdue, M.L., Rogers, H.W., O'Callaghan, D.J., and Randall, C.C. (1974), *Virology* **61**:361.

40. Vahlne, A.G. and Blomberg, J. (1974), *J. Gen. Virol.* **22**:297.

41. Dimmock, N.J. and Watson, D.H. (1969), *J. Gen. Virol.* **5**:499.

42. Laemmli, U.K. (1970), *Nature* **227**:680.

43. Studier, F.W. (1973), *J. Mol. Biol.* **79**:237

44. Powell, K.L. and Watson, D.H. (1975), *J. Gen. Virol.* **29**:167.

45. Cassai, E.N., Sarmiento, M., and Spear, P.G. (1975), *J. Virol.* **16**:1327.

46. Strand, B.C. and Aurelian, L. (1976), *Virology* **69**:438.

47. Marsden, H.S., Crombie, I.K., and Subak-Sharpe, J.H. (1977), *J. Gen. Virol.* **31**:347.

48. Perdue, M.L., Kemp, M.C., Randall, C.C., and O'Callaghan, D.J. (1974), *Virology* **59**:201.

49. Perdue, M.L., Cohen, J.C., Kemp, M.C., Randall, C.C., and O'Callaghan, D.J. (1975), *Virology* **64**:187.

50. Dolyniuk, M., Pritchett, R., and Kieff, E. (1976), *J. Virol.* **17**:935.

51. Dolyniuk, M., Wolff, E., and Kieff, E. (1976), *J. Virol.* **18**:289.

52. Fiala, M., Honess, R.W., Heiner, D.C., Heine, J.W., Murnane, J., Wallace, R., and Guze, L.B. (1976), *J. Virol.* **19**:243.

53. Kim, K.S., Sapienza, V.J., Carp, R.I., and Moon, H.M. (1976), *J. Virol.* **17**:906.

54. Kim, K.S., Sapienza, V.J., Carp, R.I., and Moon, H.M. (1976), *J. Virol.* **20**:604.

55. Stinski, M.F. (1976), *J. Virol.* **19**:595.

56. Allen, G.P. and Bryans, J.T. (1976), *Virology* **69**:751.

57. Allen, G.P. and Randall, C.C. (1979), *Virology* **92**:252.

58. Stevely, W.S. (1975), *J. Virol.* **16**:944.

59. Gupta, P., St. Jeor, S., and Rapp, F. (1977), *J. Gen. Virol.* **34**:447.

60. Killington, R.A., Yeo, J., Honess, R.W., Watson, D.H. Halliburton, I.W., and Mumford, J. (1977), *J. Gen. Virol.* **37**:297.

61. Honess, R.W. and Watson, D.H. (1977), *J. Gen. Virol.* **37**:15.

62. Spear, P.G. and Roizman, B. (1980), in *DNA Tumor Viruses (J. Tooze, Ed.), p. 615, Cold Spring Harbor Laboratory.*

63. Spring, S.B. and Roizman, B. (1968), *J. Virol.* **2**:979.

64. Sueoka, N. (1961), *J. Mol. Biol.* **3**:31.

65. Sober, H.A. (Ed.) (1970), *Handbook of Biochemistry,* The Chemical Rubber Co., Cleveland.

66. Wagner, E.K., Roizman, B., Savage, T., Spear, P.G., Mizell, M., Durr, F.E., and Sypowicz, D. (1970), *Virology* **42**:257.

67. Collard, W., Thornton, H., Mizell, M., and Green, M. (1973), *Science* **181**:447.

68. Lee, L., Kieff, E.D., Bachenheimer, S.L., Roizman, B., Spear, P.G., Burmester, B.R., and Nazerian, K. (1971), *J. Virol.* **7**:289.

69. Ludwig, H., Haines, H.G., Biswal, N., and Benyesh-Melnick, M. (1972), *J. Gen. Virol.* **14**:111.

70. Russel, W.C. and Crawford, L.V. (1964), *Virology* **26**:394.

71. Soehner, R.L., Gentry, C.A., and Randall, C.C. (1965), *Virology* **26**:394.

72. Ludwig, H. (1972), *Med. Microbiol. Immunol.* **157**:186.

73. Gravell, M. (1971), *Virology* **43**:730.

74. Plummer, G., Goodheart, C.R., Henson, D., and Bowling, C.P. (1969), *Virology* **39**:134.

75. Crawford, L.V. and Lee, A.J. (1964), *Virology* **23**:105.

76. Weinberg, A. and Becker, Y. (1969), *Virology* **39**:312.
77. Schulte-Holthausen, H. and Zur Hausen, H. (1970), *Virology* **40**:776.
78. Nonoyama, M. and Pagano, J.S. (1971), *Nat. New. Biol.* **233**:103.
79. Ludwig, H., Biswal, N., and Benyesh-Melnick, M. (1971), *Biochim. Biophys. Acta* **232**:261.
80. Jehn, U., Lindahl, T., and Klein, G. (1972), *J. Gen. Virol.* **16**:409.
81. Pritchett, R.F., Hayward, S.D., and Kieff, E.D. (1975), *J. Virol.* **15**:556.
82. Ludwig, H., Biswal, N., Bryans, J.T., and McCombs, R.M. (1971), *Virology* **45**:534.
83. Goodheart, C.R., Plummer, G., and Waner, J.L. (1968), *Virology* **35**:473.
84. Kieff, E.D., Bachenheimer, S.L., and Roizman, B. (1971), *J. Virol.* **8**:125.
85. Graham, B.J., Ludwig, H., Bronson, D.L., Benyesh-Melnick, M., and Biswal, N. (1972), *Biochim. Biophys. Acta* **259**:13.
86. Kaplan, A.S. and Ben-Porat, T. (1964), *Virology* **23**:90.
87. Goodheart, C.R. and Plummer, G. (1974), in *Progress in Medical Virology* (J.L. Melnick, Ed.), Vol. 19, S. Karger, Basel.
88. Martin, W.B., Hay, D., Crawford, L.V., LeBouvier, G.L., and Crawford, E.M. (1966), *J. Gen. Microbiol.* **45**:325.
89. Mosmann, T.R. and Hudson, J.B. (1973), *Virology* **54**:135.
90. Plummer, G., Goodheart, C.R., and Studdert, M.J. (1973), *Infect. Immun.* **8**:621.
91. Lee, L.F., Armstrong, R.L., and Nazerian, K. (1972), *Avian Dis.* **16**:799.
92. Fleckenstein, B. and Wolf, H. (1974), *Virology* **58**:55.
93. Fleckenstein, B., Bornkamm, G.W., Mulder, C., Werner, F.J., Daniel, M.D., Falk, L.A., and Delius, H. (1978), *J. Virol.* **25**:361.
94. Wagner, E.K., Tewari, K.K., Kolodner, R., and Warner, R.C. (1974), *Virology* **57**:436.
95. Skare, J., Summers, W.P., and Summers, W.C. (1975), *J. Virol.* **15**:726.
96. Hirsch, T., Reischig, J., Roubal, J., and Vonka, V. (1975), *Virology* **38**:42.
97. Walboomers, J.M.M. and Schegget, J.T. (1976), *Virology* **74**:256.

98. Pignatti, P.F., Cassai, E., Meneguzzi, G., Chenciner, N., and Milanesi, G. (1979), *Virology* **93**:260.

99. O'Callaghan, D.J., Cheevers, W.P., Gentry, G.A., and Randall, C.C. (1968), *Virology* **36**:104.

100. Rubenstein, A.S. and Kaplan, A.S. (1975), *Virology* **66**:385.

101. Darai, G., Matz, B., Schröder, C.H., Flügel, R.M., Berger, U., and Munk, K. (1979), *J. Gen. Virol.* **43**:541.

102. Chousterman, S., Lacasa, M., and Sheldrick, P. (1979), *J. Virol.* **31**:73.

103. Becker, Y., Dym, H., and Sarov, I. (1968), *Virology* **36**:184.

104. Grafstrom, R.H., Alwine, J.C., Steinhart, W.L., Hill, C.W., and Hyman, R.W. (1975), *Virology* **67**:144.

105. Wadsworth, S., Jacob, R.J., and Roizman, B. (1975), *J. Virol.* **15**:1487.

106. Frenkel, N. and Roizman, B. (1971), *J. Virol.* **8**:591.

107. Hayward, G.S., Frenkel, N., and Roizman, B. (1975), *Proc. Natl. Acad. Sci. USA* **72**:1768.

108. Hayward, G.S., Jacob, R.J., Wadsworth, S.C., and Roizman, B. (1975), *Proc. Natl. Acad. Sci. USA* **72**:4243.

109. Buchman, T.G., Roizman, B., Adams, G., and Storer, H. (1978), *J. Infect. Dis.* **138**:488.

110. Sheldrick, P. and Berthelot, N. (1974), *Cold Spring Harbor Symp. Quant. Biol.* **39**:667.

111. Wadsworth, S., Hayward, G.S., and Roizman, B. (1976), *J. Virol.* **17**:503.

112. Wagner, M.M. and Summers, W.C. (1978), *J. Virol.* **27**:374.

113. Clements, J.B., Cortini, R., and Wilkie, N.M. (1976), *J. Gen. Virol.* **30**:243.

114. Wilkie, N.M. (1976), *J. Virol.* **20**:222.

115. Wilkie, N.M. and Cortini, R. (1976), *J. Virol.* **20**:211.

116. Skare, J. and Summers, W.C. (1977), *Virology* **76**:581.

117. Delius, H. and Clements, J.B. (1976), *J. Gen. Virol.* **33**:125.

118. Roizman, B., Hayward, G., Jacob, R., Wadsworth, S.W., and Honess, R.W. (1974), *Excerpta Med. Int. Congr. Ser.* No. 350, **2**:188.

119. Jones, P.C., Hayward, G.S., and Roizman, B. (1977), *J. Virol.* **21**:268.

120. Jondal, M. and Klein, G. (1973), *J. Exp. Med.* **138**:1365.

121. Pattengale, P.K., Smith, R.W., and Gerber, P. (1973), *Lancet* **2**:93.

122. Given, D. and Kieff, E. (1978), *J. Virol.* **28**:524.

123. Hayward, S.D. and Kieff, E. (1977), *J. Virol.* **23**:421.

124. Pizzo, P.A., Magrath, I.T., Chattopadhyay, S.K., Biggar, R.J., and Gerber, P. (1978), *Nature* **272**:629.

125. Raab-Traub, N., Prichett, R., and Kieff, E. (1978), *J. Virol.* **27**:388.

126. Rymo, L. and Forsblum, S. (1978), *Nucleic Acids Res.* **5**:1387.

127. Sugden, B., Summers, W.C., and Klein, G. (1976), *J. Virol.* **18**:765.

128. Miller, G. and Lipman, M. (1973), *Proc. Natl. Acad. Sci. USA* **70**:190.

129. Miller, G., Shope, T., Lisco, H., Still, D., and Lipman, M. (1972), *Proc. Natl. Acad. Sci. USA* **69**:383.

130. Hinuma, Y., Konn, M., Yamaguchi, J., Wudarski, D.J., Blakeslee, J.R., Jr., Grace, J.T., Jr. (1967), *J. Virol.* **1**:1045.

131. Adams, A., Bjursell, G., Kaschka-Dierich, C., and Lindahl, T. (1977), *J. Virol.* **22**:373.

132. Kieff, E. and Levine, J. (1974), *Proc. Natl. Acad. Sci.* **71**:355.

133. Sugden, B. (1977), *Proc. Natl. Acad. Sci. USA* **74**:4651.

134. Hayward, S.D., Nogee, L, and Hayward, G.S. (1980), *J. Virol.* **33**:507.

135. Given, D. and Kieff, E. (1979), *J. Virol.* **31**:315.

136. Given, D., Yee, D., Griem, K., and Kieff, E. (1979), *J. Virol.* **30**:852

137. Adams, A. and Lindahl, T. (1975), *Proc. Natl. Acad. Sci. USA* **72**:1477.

138. Lindahl, T., Adams, A., Bjursell, G., Bornkamm, G.W., Kaschka-Dierich, C., and Jehn, U. (1976), *J. Mol. Biol.* **102**:511.

139. Kieff, E., Given, D., Thomas-Powell, A.L., King W., Dambaugh, T., and Raab-Traub, N. (1979), *Biochim. Biophys. Acta* **560**:355.

140. Buchman, T.G. and Roizman, B. (1978), *J. Virol.* **25**:395.

141. Buchman, T.G. and Roizman, B. (1978), *J. Virol.* **27**:239.

142. Stevely, W.S. (1977), *J. Virol.* **22**:232.

143. Ben-Porat, T., Rixon, F.J., and Blankenship, M.L. (1978), *Virology* **95**:285.

144. Deinhardt, F. (1973), in *The Herpes Viruses* (A.S. Kaplan, Ed.), p. 595, Academic Press, New York.

145. Meléndez, L.V., Hunt, R.D., King, N.W., Barahona, H.H., Daniel, M.D., Fraser, C.E.O., and Garcia, F.G. (1972), *Nature New Biol.* **235**:182.

146. Falk, L.A., Wolfe, L.G., and Deinhardt, F. (1973), *J. Natl. Cancer Inst.* **51**:165.

147. Hunt, R.D., Meléndez, L.V., King, N.W., Gilmore, C.E., Daniel, M.D., Williamson, M.E., and Jones, T.C. (1970), *J. Natl. Cancer Inst.* **44**:447.

148. Laufs, R. and Fleckenstein, B. (1972), *Med. Microbiol. Immunol.* **158**:135.

149. Laufs, R. and Fleckenstein, B. (1973), *Med. Microbiol. Immunol.* **158**:227.

150. Meléndez, L.V., Daniel, M.S., Hunt, R.D., and Garcia, F.G. (1968), *Lab. Anim. Care* **18**:374.

151. Meléndez, L.V., Hunt, R.D., Daniel, M.D., Garcia, F.G., and Fraser, C.E.O. (1969), *Lab. Anim. Care* **19**:379.

152. Bornkamm, G.W., Delius, H., Fleckenstein, B., Werner, F.J., and Mulder, C. (1976), *J. Virol.* **19**:154.

153. Fleckenstein, B., Bornkamm, G.W., and Ludwig, H. (1975), *J. Virol.* **15**:398.

154. Fijan, N.N., Wellborn, I.L., Jr., and Naftel, J.P. (1970), U.S. Dept. Interior, Bureau Sport Fish Wildlife Tech. Paper No. 43.

155. Nonoyama, M. and Pagano, J.S. (1972), *Nature New Biol.* **238**:169.

156. Frenkel, N. and Roizman, B. (1972), *J. Virol.* **10**:565.

157. Wilkie, N.M. (1973), *J. Gen. Virol.* **21**:453.

158. Ben-Porat, T., Kaplan, A.S., Stehn, B., and Rubenstein, A.S. (1976), *Virology* **69**:547.

159. Hyman, R.W., Oakes, J.E., and Kudler, L. (1977), *Virology* **76**:286.

160. Kaplan, A.S. (1969), in *Virology Monographs,* Springer Verlag, New York.

161. Roizman, B. (1969), in *Current Topics in Microbiology and Immunology,* Vol. 49. pp.1–79, Springer Verlag, Heidelberg.

162. Roizman, B. and Kieff, E.D. (1975), in *Cancer Vol. 2, A Comprehensive Treatise* (F.F. Becker, Ed.), p. 241, Plenum Press, New York.

163. Metz, D.H. (1975), *Control Processes in Virus Multiplication.* Twenty-fifth Symposium of the Society for General Microbiology, pp. 323–355, Cambridge University Press, London, New York, Melbourne.

164. Rakusanova, T., Ben-Porat, T., and Kaplan, A.S. (1972), *Virology* **49**:537.

165. Bell, D., Wilkie, N.M., and Subak-Sharpe, J.H. (1971), *J. Gen. Virol.* **13**:463.

166. Hay, J., Icoteles, G.J., Keir, H.M., and Subak-Sharpe, J.H. (1966), *Nature* **210**:387.

167. Wagner, E.K. and Roizman, B. (1969), *J. Virol.* **4**:36.

168. Hay, J. and Low, M. (1970), *Biochem. J.* **117**:21.

169. Rakusanova, T. and Kaplan, A.S. (1970), *Fed. Proc. Fed. Amer. Soc. Exp. Biol.* **29**:309.

170. Aurelian, L. and Roizman, B. (1964), *Virology* **22**:452.

171. Roizman, B. and Roane, P.R., Jr. (1964), *Virology* **22**:262.

172. Gergely, L., Klein, G., and Ernberg, I. (1971), *Int. J. Cancer* **7**:293.

173. Ben-Porat, T. and Kaplan, A.S. (1965), *Virology* **25**:22.

174. Halliburton, I.W. and Timbury, M.C. (1976), *J. Gen. Virol.* **30**:207.

175. Ben-Porat, T. and Kaplan, A.S. (1973), in *The Herpes Viruses* (A.S., Kaplan, Ed.), pp. 163–220, Academic Press, New York.

176. Fenwick, M.L. and Walker, M.J. (1978), *J. Gen. Virol.* **41**:37.

177. Roizman, B., Borman, G.S., and Kamali-Rousta, M. (1965), *Nature* **206**:1374.

178. Sydiskis, R.J. and Roizman, B. (1966), *Science* **153**:76.

179. Ben-Porat, T., Rakusanova, T., and Kaplan, A.S. (1971), *Virology* **46**:890.

180. Powell, K.L. and Courtney, R.J. (1975), *Virology* **66**:217.

181. Pereira, L., Wolff, M.H., Fenwick, M.L., and Roizman, B. (1977), *Virology* **77**:733.

182. Fenwick, M., Morse, L.S., and Roizman, B. (1979), *J. Virol.* **29**:825.

183. Roizman, B. (1963), in *Viruses, Nucleic Acids and Cancer,* Proceedings of the 17th Annual Symposium, M.D. Anderson Hospital and Tumor Institute, pp. 205–223. Williams & Wilkins, Baltimore.

184. Sauer, G., Orth, H.D., and Munk, K. (1966), *Biochim. Biophys. Acta* **119**:331.

185. Sauer, G. and Munk, K. (1966), *Biochim. Biophys. Acta* **119**:341.

186. Flanagan, J.F. (1967), *J. Virol.* **1**:583.

187. Ward, R.L. and Stevens, J.G. (1975), *J. Virol.* **15**:71.

188. Silverstein, S., Millette, R., Jones, P.C., and Roizman, B. (1976), *J. Virol.* **18**:977.

189. Jamieson, A.T., MacNab, J.C.M., Perbal, B., and Clements, J.B. (1976), *J. Gen. Virol.* **32**:493.

190. Stringer, J.R., Holland, L.E., Swanstrom, R.I., Pivo, K., and Wagner, E.K. (1977), *J. Virol.* **21**:889.

191. Brawerman, G., Mendecki, J., and Lee, S.Y. (1972), *Biochemistry* **11**:637.

192. Zimmerman, S.B. and Sandeen, G. (1966), *Anal. Biochem.* **14**:269.

193. Watson, R.J. and Clements, J.B. (1978), *Virology* **91**:364.

194. Clements, J.B., Watson, R.J., and Wilkie, N.M. (1977), *Cell* **12**:275.

195. Szybalski, W. (1968) in *Methods in Enzymology* (L. Grossman and K. Moldave, Eds.) Vol. 12B, pp. 330–360, Academic Press, New York.

196. Jones, P.C. and Roizman, B. (1979), *J. Virol.* **31**:299.

197. Lozeron, H.A. and Szybalski, W. (1966), *Biochem. Biophys. Res. Commun.* **23**:612.

198. Lando, D.L. and Ryhiner, M.L. (1969), *C.R. Acad. Sci. (Paris)* **269**:527.

199. Graham, F.L., Veldhuisen, G., and Wilkie, N.M. (1973), *Nature New Biol.* **245**:265.

200. Sheldrick, P., Laithier, M., Lando, D.L., and Ryhiner, M.L. (1973), *Proc. Natl. Acad. Sci. USA* **70**:3621.

201. Lindell, T.J., Weinberg, F., Morris, P.W., Roeder, R.G., and Rutter, W.J. (1970), *Science* **170**:447.

202. Chambon, P., Gissinger, F., Mandel, J.L., Kedinger, C., Gniazdowski, M., and Meihlac, M. (1970), *Cold Spring Harbor Symp. Quant. Biol.* **35**:693.

203. Jacob, S.T., Sajdel, E.M., Muecke, W., and Munro, H.N. (1970), *Cold Spring Harbor Symp. Quant. Biol.* **35**:681.

204. Constanzo, F., Campadelli-Fuime, G., Foa-Tomasi, L., and Cassai, E. (1977), *J. Virol.* **21**:996.

205. Ben-Zeev, A. and Becker, Y. (1977), *Virology* **76**:246.

206. Alwine, J.C., Steinhart, W.H., and Hill, C.W. (1974), *Virology* **60**:302.

207. Wagner, E.K. (1972), *Virology* **47**:502.

208. Wagner, E.K., Swanstrom, R.I., and Stafford, M.G. (1972), *J. Virol.* **10**:675.

209. Swanstrom, R.I. and Wagner, E.K. (1974), *Virology* **60**:522.

210. Swanstrom, R.I., Pivo, K., and Wagner, E.K. (1975), *Virology* **66**:140.

211. Clements, J.B. and Hay, J. (1977), *J. Gen. Virol.* **35**:1.

212. Preston, C.M. (1979), *J. Virol.* **29**:275.

213. Preston, C.M. (1979), *J. Virol.* **32**:357.

214. Bechet, J.M. and Montagnier, L. (1975), *C.R. Acad. Sci. (Paris)* **280**:217.

215. Ben-Zeev, A. and Becker, Y. (1975), *Nature* **254**:719.

216. Kozak, M. and Roizman, B. (1974), *Proc. Natl. Acad. Sci.* **71**:4322.

217. Jacquemont, B. and Roizman, B. (1975), *J. Virol.* **15**:707.

218. Frenkel, N. and Roizman, B. (1972), *Proc. Natl. Acad. Sci. USA* **68**:2654.

219. Silverstein, S., Bachenheimer, S.L., Frenkel, N., and Roizman, B. (1973), *Proc. Natl. Acad. Sci. USA* **70**:2101.

220. Wagner, E.K. and Roizman, B. (1969), *Proc. Natl. Acad. Sci. USA* **64**:626.

221. Roizman, B., Bachenheimer, S.L., Wagner, E.K., and Savage, T. (1970), *Cold Spring Harbor Symp. Quant. Biol.* **35**:753.

222. Bachenheimer, S.L. and Roizman, B. (1972), *J. Virol.* **10**:875.

223. Bartkosi, M. and Roizman, B. (1976), *J. Virol.* **20**:583.

224. Moss, B., Gershowitz, A., Stringer, J.R., Holland, L.E., and Wagner, E.K. (1977), *J. Virol.* **23**:234.

225. Adams, J. and Cory, S. (1975), *Nature* **255**:28.

226. Desrosiers, R., Frederici, K., and Rottman, F. (1974), *Proc. Natl. Acad. Sci. USA* **71**:3971.

227. Perry, R.P. and Kelley, D. (1974), *Cell* **1**:37.

228. Lavi, S. and Shatkin, A.J. (1975), *Proc. Natl. Acad. Sci. USA* **72**:2012.

229. Shatkin, A.J. (1974), *Proc. Natl. Acad. Sci. USA* **71**:3204.

230. Wei, C.M. and Moss. B. (1974), *Proc. Natl. Acad. Sci. USA* **71**:3014.

231. Frenkel, N., Silverstein, S., Cassai, E., and Roizman, B. (1973), *J. Virol.* **11**:886.

232. Leung, W.C. (1978), *J. Virol.* **27**:269.

233. Preston. V.G. (1981), *J. Virol.* **39**:150.

234. Watson, R.J., Preston, C.M., and Clements, J.B. (1979), *J. Virol.* **31**:42.

235. Clements, J.B., McLauchlan, J., and McGeoch, D.J. (1979), *Nucleic Acids Res.* **7**:77.

236. Millette, R. and Talley Brown, S. (1980), Abstract 723 of the 9th Annual ICN-UCLA Symposia Molecular and Cellular Biology, Keystone, Colorado.

237. Post, L.E., Mackem, S., and Roizman, B. (1981), *Cell* **24**:555.

238. Stringer, J.R., Holland, L.E., and Wagner, E.K. (1978), *J. Virol.* **27**:56.

239. Bailey, J.M. and Davidson, N. (1976), *Anal. Biochem.* **70**:75.

240. Holland, L.E., Anderson, K.P., Stringer, J.R., and Wagner, E.K. (1979), *J. Virol.* **31**:447.

241. Bone, D.R. and Courtney, R.J. (1974), *J. Gen. Virol.* **24**:17.

242. Honess, R.W. and Roizman, B. (1974), *J. Virol.* **14**:8.

243. Honess, R.W. and Watson, D.H. (1977), *J. Virol.* **21**:584.

244. Powell, K.L., Purifoy, D.J.M., and Courtney, R.J. (1975), *Biochem. Biophys. Res. Commun.* **66**:262.

245. Anderson, K.P., Stringer, J.R., Holland, L.E., and Wagner, E.K. (1979), *J. Virol.* **30**:805.

246. Honess, R.W. and Roizman, B. (1973), *J. Virol.* **12**:1347.

247. Anderson, K.P., Frink, R.J., Devi, G.B., Gaylord, B.H., Costa, R.H., and Wagner, E.K. (1981), *J. Virol.* **37**:1011.

248. Costa, R.H., Devi, G.B., Anderson, K.P., Gaylord, B.H., and Wagner, E.K. (1981), *J. Virol.* **38**:483.

249. Strauss, G.P., Maichle, I.B., Schatten, R., and Kaerner, H.C. (1977), *Nucleic Acids Res.* **4**:1793.

250. Chenciner, N. (1980), *These de doctorat de 3ᵉ cycle.* Université Paris VI.

251. Feldmann, L., Rixon, F.J., Jean, J.H., Ben-Porat, T., and Kaplan, A.S. (1979), *Virology* **97**:316.

252. Rakusanova, T., Ben-Porat, T., Himeno, M., and Kaplan, A.S. (1971), *Virology* **46**:877.

253. Tracy, S. and Desrosiers, R.C. (1980), *Virology* **100**:204.

254. Hayward, D. and Kieff, E. (1976), *J. Virol.* **18**:518.

255. Orellana, T. and Kieff, E. (1977), *J. Virol.* **22**:321.

256. Powell, A.L.T., King, W., and Kieff, E. (1979), *J. Virol.* **29**:261.

257. Schneweiss, E.K. (1962), *Z. Immunitatsforsch.* **124**:24.

258. Newton, A.A. and Stocker, M.G.P. (1958), *Virology* **5**:549.

259. Munk, K. and Sauer, G. (1963), *Z. Naturforsch.* B **18**:211.

260. Munk, K. and Sauer, G. (1964), *Virology* **22**:153.

261. Cohen, G.H., Vaughan, R.K., and Lawrence, W.C. (1971), *J. Virol.* **7**:783.

262. St. Jeor, S.C. and Hutt, R. (1977), *J. Gen. Virol.* **37**:65.

263. Muller, M.T. and Hudson, J.B. (1977), *J. Virol.* **22**:267.

264. Lawrence, W.C. (1971), *Am. J. Vet. Res.* **32**:41.

265. Lawrence, W.C. (1971), *J. Virol.* **7**:736.

266. O'Callaghan, R., Randall, C.C., and Gentry, G.A. (1972), *Virology* **49**:784

267. Kaplan, A.S. and Ben-Porat, T. (1963), *Virology* **19**:205.

268. Kaplan, A.S., Ben-Porat, T. and Coto, C. (1967), in *Molecular Biology of Viruses* (J. Colter, Ed.), pp. 527–545, Academic Press, New York.

269. Keir, H.M. and Gold, E. (1963), *Biochim. Biophys. Acta* **72**:263.

270. Russel, W.C., Gold, E., Keir, H.M., Omura, H., Watson, D.H., and Wildy, P. (1964), *Virology* **22**:103.

271. McAuslan, B.R., Herde, P., Pett, D., and Ross, J. (1965), *Biochem. Biophys. Res. Comm.* **20**:586.

272. Morrison, J.M. and Keir, H.M. (1966), *Biochem. J.* **98**:37C.

273. Morrison, J.M. and Keir, H.M. (1967), *Biochem. J.* **103**:70.

274. Morrison, J.M. and Keir, H.M. (1968), *Biochem. J.* **110**:39P.

275. Morrison, J.M. and Keir, H.M. (1968), *J. Gen. Virol.* **3**:337.

276. Klemperer, H.G., Haynes, G.R., Shedden, W.I.H., and Watson, D.H. (1967), *Virology* **31**:120.

277. Hamada, C., Kamiya, T., and Kaplan, A.S. (1966), *Virology* **28**:271.

278. Kit, S., Dubbs, D.R., and Anken, M. (1967), *J. Virol.* **1**:238.
279. Buchan, A. and Watson, D.H. (1969), *J. Gen. Virol.* **4**:461.
280. Hay, J., Perera, P.A., Morrison, J.M., Gentry, G.A., and Subak-Sharpe, J.H. (1971), in *Ciba Symposium on Strategy of the Viral Genome* (G.E.W. Wolstenholme and M. O'Connor, Eds.), Churchill Livingstone, London.
281. Keir, H.M. (1968), in *Molecular Biology of Viruses,* Vol. 18, pp. 67–99, Cambridge University Press, Cambridge.
282. Chan, T. (1977), *Proc. Natl. Acad. Sci. USA* **74**:1734.
283. Reichard, P., Canellakis, Z.N., and Canellakis, E.S. (1960), *Biochim. Biophys. Acta* **41**:558.
284. Reichard, P., Canellakis, Z.N., and Canellakis, E.S. (1961), *J. Biol. Chem.* **236**:2514.
285. Cohen, G.H. (1972), *J. Virol.* **9**:408.
286. Keir, H.M. (1965), *Prog. Nucleic Acid Res. Mol. Biol.* **4**:81.
287. Keir, H.M., Subak-Sharpe, J.H., Shedden, W.I.H., Watson, D.H., and Wildy, P. (1966), *Virology* **30**:154.
288. Keir, H.M., Hay, J., Morrison, J.M., and Subak-Sharpe, J.H. (1966), *Nature* **210**:369.
289. Shedden, W.I.H., Subak-Sharpe, J.H., Watson, D.H., and Wildy, P. (1966), *Virology* **30**:154.
290. Weissbach, A., Schlabach, A., Fridlender, B., and Bolden, A. (1971), *Nature New Biol.* **231**:167.
291. Halliburton, I.W. and Andrew, J.C. (1976), *J. Gen. Virol.* **30**:145.
292. Boezi, J.A., Lee, L.F., Blakesley, R.W., Koenig, M., and Towle, H.C. (1974), *J. Virol.* **14**:1209.
293. Hirai, K., Furukawa, T., and Plotkin, S. (1976), *Virology* **70**:251.
294. Huang, E.S. (1975), *J. Virol.* **16**:298.
295. Stevens, J.G. and Jackson, N.L. (1967), *Virology* **32**:654.
296. Cohen, J.C., Perdue, M.L., Randall, C.C., and O'Callaghan, D.J. (1975), *Virology* **67**:56.
297. Kemp, M.C., Cohen, J.C., O'Callaghan, D.J., and Randall, C.C. (1975), *Virology* **68**:467.
298. Shipkowitz, N.L., Bower, R.R., Appell, R.N., Nordeen, C.W., Overby, L.R., Roderick, W.R., Schleicher, J.B., and Van Esch, A.M. (1973), *Appl. Microbiol.* **27**:264.
299. Overby, L.R., Robishaw, E.E., Schliecher, J.B., Rueter, A., Shipkowitz, N.L., and Mao, J.C.H. (1974), *Antimicrob. Agents Chemother.* **6**:360.

300. Mao, J.C.-H., Robishaw, E.E., and Overby, L.R. (1975), *J. Virol.* **15**:1281.
301. Mao, J.C.-H. and Robishaw, E.E. (1975), *Biochemistry* **14**:5475.
302. Huang, E.S. (1975), *J. Virol.* **16**:1560.
303. Bolden, A., Aucker, J., and Weissbach, A. (1975), *J. Virol.* **16**:1584.
304. Huang, E.S., Huang, C.H., Huong, S.M., and Selgrade, M. (1976), *Yale J. Biol. Med.* **49**:93.
305. Leinbach, S.S., Reno, J.M., Lee, L.F., Isbell, A.F., and Boezi, J.A. (1976), *Biochemistry* **15**:426.
306. Lee, L.F., Nazerian, K., Leinbach, S.S., Reno, J.M., and Boezi, J.A. (1976), *J. Natl. Cancer Inst.* **56**:823.
307. Nazerian, K. and Lee, L.F. (1976), *Virology* **74**:188.
308. Yajimi, Y., Tanaka, A., and Nonoyama, M. (1976), *Virology* **71**:352.
309. Allen, G.P., O'Callaghan, D.J., and Randall, C.C. (1977), *Virology* **76**:395.
310. Summers, W.C. and Klein, G. (1976), *J. Virol.* **18**:151.
311. Elliott, R.M., Bateson, A., and Kelly, D.C. (1980), *J. Virol.* **33**:539.
312. Hay, J. and Subak-Sharpe, J.H. (1976), *J. Gen. Virol.* **31**:145.
313. Joffre, J.T., Schaffer, P.A., and Parris, D.S. (1977), *J. Virol.* **23**:833.
314. Chen, M.S., Summers, W.P., Walker, J., Summers, W.C., and Prusoff, W.H. (1979), *J. Virol.* **30**:942.
315. Purifoy, D.J.M. and Benyesh-Melnick, M. (1975), *Virology* **68**:374.
316. Hay, J., Moss, H., Jamieson, A.T., and Timbury, M.C. (1976), *J. Gen. Virol.* **31**:65.
317. Bone, D.R., Brown, M.S., Crombie, I., and Francke, B. (1978), *J. Virol.* **28**:14.
318. Adler, R., Glorioso, J.C., and Levine, M. (1978), *J. Gen. Virol.* **39**:9.
319. Campbell, W.F., Murray, B.K., Biswal, N., and Benyesh-Melnick, M. (1974), *J. Natl. Cancer Inst.* **52**:757.
320. Geder, L., Vaczi, L., and Jeney, E. (1971), *Acta Virologica* **15**:35.
321. Tucker, A. and Docherty, J. (1975), *Infect. Immun.* **11**:556.
322. Biegeleisen, K. and Rush, M.G. (1976), *Virology* **69**:246.

323. Biegeleisen, K., Yanagi, K., and Rush, M.G. (1977), *Virology* **83**:221.

324. Hoggan, M.D., Roizman, B., and Turner, T.B. (1960), *J. Immunol.* **84**:152.

325. Biswal, N., Murray, B.K., and Benyesh-Melnick, M. (1974), *Virology* **61**:87.

326. Ben-Porat, T. and Kaplan, A.S. (1963), *Virology* **20**:310.

327. Kobler, A.R. (1975), *J. Virol.* **15**:322.

328. Ben-Porat. T., Stehn, B., and Kaplan, A.S. (1976), *Virology* **71**:412.

329. Ben-Porat, T., Blankenship, M.L., De Marchi, J., and Kaplan, A.S. (1977), *J. Virol.* **22**:734.

330. Schlomai, J., Friedman, A., and Becker, Y. (1976), *Virology* **69**:647.

331. Pignati, P.F., Cassai, E., and Bertazzoni U. (1979), *J. Virol.* **32**:1033.

332. Nathans, D. and Danna, K.J. (1972) *Nature New Biol.* **236**:200.

333. Hirsch, I., Cabral, G., Patterson, M., and Biswal, N. (1977), *Virology* **81**:48.

334. Ben-Zeev, A., Weinberg, E., and Becker, Y. (1974), *J. Gen. Virol.* **25**:63.

335. Jean, J.H. and Ben-Porat, T. (1976), *Proc. Natl. Acad. Sci.* **73**:2674.

336. Friedmann, A. and Becker, Y. (1977), *J. Gen. Virol.* **37**:205.

337. Jacob, R.J. and Roizman, B. (1977), *J. Virol.* **23**:394.

338. Jacob, R.J., Morse, L.S., and Roizman, B. (1979), *J. Virol.* **29**:448.

339. Gilbert, W. and Dressler D. (1968), *Cold Spring Harbor Symp. Quant. Biol.* **33**:473.

340. Roizman, B., Hayward, G., Jacob, R., Wadsworth, S.W., Frenkel, N., Honess, R.W., and Kozak, M. (1975), *Proceedings of the Symposium on Herpes Viruses and Oncogenesis, Nuremberg* (G. de The, M.A. Epstein, and H. Zur Hausen, Eds.), p. 3, IARC, Lyon.

341. Fujiwara, S. and Kaplan, A.S. (1967), *Virology* **32**:60.

342. Sydiskis, R.J. and Roizman, B. (1967), *Virology* **32**:678.

343. Watson, D.H., Shedden, W.I.H., Elliot, A., Tetsuka, T., Wildy, P., Bourgaux-Ramoisy, D., and Gold, E. (1966), *Immunology* **11**:399.

344. Watson, D.H. and Wildy, P. (1969), *J. Gen. Virol.* **4**:163.

345. Honess, R.W. and Watson, D.H. (1974), *J. Gen. Virol.* **22**:171.

346. Courtney, R.J. and Powell, K.L. (1974), *Proceedings of the Symposium on Herpes Viruses and Oncogenesis* (G. de The, M.A. Epstein, and H. Zur Hausen, Eds.) p. 63–73, IARC, Lyon.

347. Spear, P.G., Keller, J.M., and Roizman, B. (1970), *J. Virol.* **5**:123.

348. Savage, T., Roizman, B., and Heine, J.W. (1972), *J. Gen. Virol.* **17**:31.

349. Heine, J.W., Spear, P.G., and Roizman, B. (1972), *J. Virol.* **9**:431.

350. Spear, P.G. (1976), *J. Virol.* **17**:991.

351. Stinsky, M.F. (1977), *J. Virol.* **23**:751.

352. Ben-Porat, T., Shimono, H., and Kaplan, A.S. (1970), *Virology* **41**:256.

353. Ben-Porat, T. and Kaplan, A.S. (1970), *Virology* **41**:265.

354. Kaplan, A.S. and Ben-Porat, T. (1970), *Proc. Natl. Acad. Sci. USA* **66**:799.

355. Kaplan, A.S. (1972), *Am. J. Clin. Pathol.* **57**:783.

356. Erickson, J.S. and Kaplan, A.S. (1973), *Virology* **55**:94.

357. Kaplan, A.S. and Ben-Porat, T. (1976), *Virology* **70**:561.

358. Gibson, W. and Roizman, B. (1974), *J. Virol.* **13**:155.

359. Marsden, H.S., Stow, N.D., Preston,V.G., Timbury, M.C., and Wilkie, N.M. (1978), *J. Virol.* **20**:624.

360. Wilcox, K.W., Kohn, A., Sklyanskaya, E., and Roizman, B. (1980), *J. Virol.* **33**:167.

361. Roizman, B., Kozak, M., Honess, R.W., and Hayward, G. (1974), *Cold Spring Harb. Symp. Quant. Biol.* **39**:687.

362. Bookout, J.B. and Levy, C.C. (1980), *Virology* **101**:198.

363. Honess, R.W. and Roizman, B. (1975), *Proc. Natl. Acad. Sci. USA* **72**:1276.

364. Kit, S., Dubbs, D.R., and Schaffer, P.A. (1978), *Virology* **85**:456.

365. Grodzicker, T., Anderson C., Sambrook, J., and Mathews, M.B. (1977), *Virology* **80**:111.

366. Mautner, V., Williams, J., Sambrook, J., Sharp, P.A., and Grodzicker, T. (1975), *Cell* **5**:93.

367. Morse, L.S., Pereira, L., Roizman, B., and Schaffer, P.A. (1978), *J. Virol.* **26**:389.

368. Preston, V.G., Davidson, A.J., Marsden, H.S., Timbury, M.C., Subak-Sharpe, J.H., and Wilkie, N.M. (1978), *J. Virol.* **28**:499.

369. Kozak, M. and Roizman, B. (1975), *J. Virol.* **15**:36.

370. Schaffer, P.A., Vonka, V., Lewis, R., and Benyesh-Melnick, M. (1970), *Virology* **42**:1144.

371. Timbury, M.C. (1971), *J. Gen. Virol.* **13**:373.

372. Brown, S.M., Ritchie, D.A., and Subak-Sharpe, J.H. (1973), *J. Gen. Virol.* **18**:329.

373. Schaffer, P.A., Aron, G.M., Biswal, N., and Benyesh-Melnick, M. (1973), *Virology* **52**:57.

374. Manservigi, R. (1974), *Appl. Microbiol.* **27**:1034.

375. Esparza, J., Purifoy, D.J.M., Schaffer, P.A., and Benyesh-Melnick, M. (1974), *Virology* **57**:554.

376. Hughes, R.G., Jr. and Mynyon, W.H. (1975), *J. Virol.* **16**:275.

377. Schaffer, P.A. (1975), *Curr. Top. Microbiol. Immunol.* **70**:51.

378. Koment, R.W. and Rapp, F. (1975), *J. Virol.* **15**:812.

379. Summers, W.P., Wagner, M., and Summers, W.C. (1975), *Proc. Natl. Acad. Sci. USA* **72**:4081.

380. Zygraich, N. and Huygelen, C. (1973), *Arch. Gesamte Virusforsch* **43**:103.

381. Schaffer, P.A., Carter, V.C., and Timbury, M.C. (1978), *J. Virol.* **27**:490.

382. Smith, J.D., Barnett, L., Brenner, S., and Russel, R.L. (1970), *J. Mol. Biol.* **54**:1.

383. Kit, S. and Dubbs, D.R. (1963), *Biochem. Biophys. Res. Commun.* **11**:55.

384. Jamieson, A.T., Gentry, G.A., and Subak-Sharpe, J.H. (1974), *J. Gen. Virol.* **24**:465.

385. Keller, J.M., Spear, P.G., and Roizman, B. (1970), *Proc. Natl. Acad. Sci. USA* **65**:865.

386. Heine, J.W., Spear, P.G., and Roizman, B. (1972), *J. Virol.* **9**:431.

387. Manservigi, R., Spear, P.G., and Buchan, A. (1977), *Proc. Natl. Acad. Sci. USA* **74**:3913.

388. Koment, R.W. and Rapp, F. (1975), *Virology* **64**:164.

389. Westmoreland, D. and Rapp, F. (1976), *J. Virol.* **18**:92.

390. Ejercito, P.M., Kieff, E.D., and Roizman, B. (1968), *J. Gen. Virol.* **3**:357.

391. Gray, A., Tokumaru, T., and McN Scott, T.F. (1958), *Arch. Gesamte Virusforsch.* **8**:59.

392. Hoggan, M.D. and Roizman, B. (1959), *Am. J. Hyg.* **70**:208.

393. Schneweis, K.E. (1962), *Zentralbl. Bakteriol. Parasitenkd. Infektionskr. Hyg. Abt. Orig.* **186**:467.

394. Wheeler, C.E., Jr. (1964), *J. Immunol.* **93**:749.

395. Timbury, M.C., Theriault, A., and Elton, R.A. (1974), *J. Gen. Virol.* **23**:219.

396. Cassai, E., Manservigi, R., Corallini, A., and Terni, M. (1975), *Intervirology* **6**:212.

397. Purifoy, D.J.M. and Powell, K.L. (1977), *J. Virol.* **24**:470.

398. Purifoy, D.J.M., Lewis, R.B., and Powell, K.L. (1977), *Nature* **269**:621.

399. Aron, G.M., Purifoy, D.J.M., and Schaffer, P.A. (1975), *J. Virol.* **16**:498.

400. Gentry, G.A. and Aswell, J.A. (1975), *Virology* **65**:294.

401. Courtney, R.J., Schaffer, P.A., and Powell, K. (1976), *Virology* **75**:306.

402. Hershey, A.D. (1946), *Cold Spring Harb. Symp. Quant. Biol.* **11**:67.

403. Hershey, A.D. and Rotman, R. (1949), *Genetics* **34**:44.

404. Brown, S.M. (1972), Ph.D. Thesis, University of Glasgow.

405. Schaffer, P.A., Tevethia, M.J., and Benyesh-Melnick, M. (1974), *Virology* **58**:219.

406. Crombie, I.K. (1975), Ph.D. Thesis, University of Glasgow.

407. Benyesh-Melnick, M., Schaffer, P.A., Courtney, R.J., Esparaza, J., and Kimura, S. (1975), *Cold Spring Harbor Symp. Quant. Biol.* **39**:731.

408. Timbury, M.C. and Calder, L. (1976), *J. Gen. Virol.* **30**:179.

409. Ritchie, D.A. (1973), *Br. Med. Bull.* **29**:247.

410. Timbury, M.C., Hendricks, M.L., and Schaffer, P.A. (1976), *J. Virol.* **18**:1139.

411. Timbury, M.C. and Subak-Sharpe, J.H. (1973), *J. Gen. Virol.* **18**:347.

412. Messer, L.I. and Timbury, M.C. (1979), *J. Gen. Virol.* **45**:431.

413. Lai, C.-J. and Nathans, D. (1974), *Virology* **60**:466.

414. Lai, C.-J. and Nathans, D. (1975), *Virology* **66**:70.

415. Mantei, N., Boyer, H.W., and Goodman, H.M. (1975), *J. Virol.* **16**:754.

416. Miller, L.K. and Fried, M. (1976), *J. Virol.* **18**:824.

417. Feunteun, J., Sampayrac, L., Fluck, M., and Benjamin, T. (1976), *Proc. Natl. Acad. Sci. USA* **73**:4169.

418. Weisbeek, P.J., Vereyken, J.M., Baas, P.D., Janez, H.S., and Van Arkel, G.A. (1976), *Virology* **72**:61.

419. Graham, F.L. and Van Der Eb, A.J. (1973), *Virology* **52**:456.

420. Wilkie, N.M., Clements, J.B., NacNab, J.C.M., and Subak-Sharpe, J.H. (1974), *Cold Spring Harbor Symp. Quant. Biol.* **39**:657.

421. Stow, N.D., Subak-Sharpe, J.H., and Wilkie, N.M. (1978), *J. Virol.* **28**:182.

422. Wigler, M., Silverstein, S., Lee, L.-S., Pellicer, A., Cheng, Y.C., and Axel, R. (1977), *Cell* **11**:223.

423. Parris, D.S., Dixon, R.A.F., and Schaffer, P.A. (1980), *Virology* **100**:275.

424. Knipe, D.M., Ruyechan, W.T., and Roizman, B. (1979), *J. Virol.* **29**:698.

425. Morse, L.S., Buchman, T.G., Roizman, B., and Schaffer, P.A. (1977), *J. Virol.* **24**:231.

426. Stow, N.D. and Wilkie, N.M. (1978), *Virology* **90**:1.

427. Chartrand, P., Stow, N.D., Timbury, M.C., and Wilkie, N.M. (1979), *J. Virol.* **31**:265.

428. Ruyechan, W.T., Morse, L.S., Knipe, D.M., and Roizman, B. (1979), *J. Virol.* **29**:677.

429. Roizman, B. (1979), *Cell* **16**:481.

ONCOGENIC TRANSFORMATION

The preceding chapter of this monograph was devoted to the interactions involved at a molecular level between mammalian cells and polyomaviruses or herpesviruses that lead to the production of infectious progeny (lytic cycle). A common feature of the polyomaviruses and herpesviruses is that in some cases they are able to interact with their host in such a way as to induce the formation of tumors. The association of these DNA tumor viruses with different kinds of cancer has been discussed in some excellent reviews (1–4). Therefore, we will focus in this chapter only on the *in vitro* transforming abilities of the herpesviruses with appropriate comparisons to polyomaviruses.

I. PROPERTIES OF TRANSFORMED CELLS

The definition of a transformed state may be given only after a comparison of the biological properties of normal and transformed cells. This, of course, implies that normal cells are available. Before analyzing the particular properties related to the transformed state, we should consider what are usually called "normal cells."

A. NORMAL CELLS

The normal cells generally used come from two different origins: primary cultures or established cell lines.

1. Primary Cultures

Primary cultures are composed of cells obtained after dissociation of mammalian tissues by mechanical and enzymatic procedures (see review, ref. 5). Such cells may be grown on a plastic or glass support in the presence of growth medium, generally supplemented with 5 to 10% calf or human serum. A fundamental characteristic of these cells is that they form monolayers of parallel cells on the support. This is the result of a firm adhesion to the substratum and of contact inhibition of movement (6,7). Under such conditions, cells are generally spread and flattened, and following a collision they turn aside and move in another direction. Frequently, these growth conditions will favor the multiplication of fibroblastic cells, and after two or three passages the proportion of epithelial cells is found to be very much reduced. Generally, the cell population will cease to grow within a few days of culture, as most of the cells rapidly die (8,9) for unknown reasons. Todaro and Green (10) established that an essential factor governing the rapidity of the de-

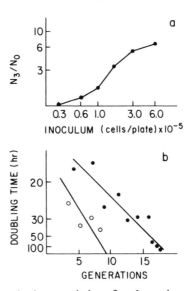

FIGURE 1.   Growth characteristics of embryonic mouse fibroblasts. a, The effect of different inoculum density on the growth of embryonic cells is measured by seeding the cultures with increasing cell densities and determining the growth rate of the cells after 3 days in culture (N_3/N_0). b, Increase of doubling time (Dt) observed with two different cultures; 0, inoculum 10^5 cells, 0, inoculum 6×10^5.

cline relies on the density of cells used to inoculate the cultures. Therefore, cells from a healthy subconfluent primary culture were split into several secondary cultures by seeding plates with varying densities of cells. After 3 days of incubation at 37°C, the number of cells was counted in each culture and the results obtained showed that a minimal seeding cell density (3×10^4 cells per 450-mm plate) was necessary before a significant growth of cell population (Fig. 1a) was observed. However, in all cases, the growth rate of the cells declined rapidly as the number of generations increased. After 10 to 20 generations, the doubling time of the cells reached values equal to or greater than 70 hours, whereas the initial doubling time was about 15 hours (Fig. 1b). The limited lifespan of these normal primary cells has led many investigators to search for cell lines that are able to grow indefinitely without being tumorigenic when injected into animals.

2. Established Cell Lines

Established cell lines were defined in the past as cell populations that are able to grow for several months under laboratory conditions (11,12), although it has been reported that cells can be grown for many generations without becoming established (9,13). The studies performed by Todaro and Green (10) allowed them to define quantitatively what should be called established cells. These authors studied the process of establishment by investigating the variation of growth rate of cultured mouse fibroblasts upon successive transfers. They showed that after a period of time corresponding to 10 to 20 generations from the beginning of the culture, the cells stopped dividing, this being shown by the ratio of output/input cells (N_2/N_0). After a time corresponding to 15 to 30 generations, the growth rate increased to reach a value similar to that observed before the "crisis" period (see Fig. 2). The cells obtained at this time were shown to be established in that their growth rate was never found to decrease, and they could now be grown for a considerable number of generations. Different transfer schedules (transfer every 3 or 6 days) led to the isolation of mouse 3T3 and 3T6 cell lines (10). Similarly, FR 3T3 rat cell lines have been obtained by this method (14) (see Fig. 2). The characteristics of these two kinds of normal cells are probably complementary. On the one hand, primary cultures are probably more likely to represent the physiological state of mammalian cells *in vivo* than are established cells which may have been selected for a "nonphysiological" property, that is, continuous growth ability. *In vivo* studies (15–20) showed that immortality is certainly not a common trait of normal mammalian cells. On the other hand, primary cultures have been shown to be a genetically heterogenous mixture of different cell types with limited growth ability as compared with established cell lines that may be purified by cloning. However, alterations of the cellular karyotype have also been found to occur during the selection of estab-

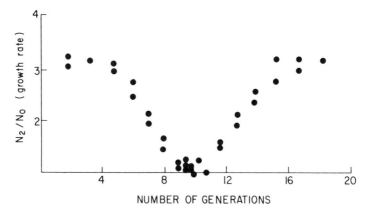

FIGURE 2. Isolation of FR3T3 rat cells. Primary rat embryo cells were seeded at a density of 3×10^5 cells per plate (N_0) and incubated for 3 days at 37°C. After 3 days, the number of cells (N_2) was determined and the operation was repeated. The growth rate N_2/N_0 is expressed as a function of number of generations at each replating. Data from Seif (14).

TABLE I.   Chromosome Alterations During the
Establishment of Mouse Cells[a]

Schedule of Transfect	Number of Generations	Percentage of Diploid Cells
3T3	2	93
	10	92
	20	79
	22	4
	25	0
3T6	11	95
	21	92
	28	89
	31	87
	35	48
	39	6

[a] Data from Todaro and Green (10).

lished cells (10,21–23) (Table I). At late times, abnormal chromosomes appear and the number of metacentric and minute chromosomes progressively increase (10). Rodent cells, and especially murine cells, have been reported to be genetically labile and to create heterogenous cell populations (24). FR 3T3 cells freshly established and recloned were also found to contain a small percentage of tetraploid cells (14). All these observations demonstrate the need for a periodic control of the percentage of polyploid cells in a given established cell line, especially when such cells have been used for long-run experiments involving many passages and have been submitted to very strong selective procedures. Slight karyotypic changes have also been reported in the case of other established cell lines from hamster or pig (25–27).

However, the biological properties (considered as normal) of such established cells and the ease with which they can be grown have been determinant factors for their extensive use in many different laboratories.

B. TRANSFORMED CELLS

When normal cells are grown under laboratory conditions, they cease to divide after they have reached a particular density corresponding to the formation of a confluent monolayer of flat cells. The saturation density of cells finally obtained is independent of the cell density used to seed the culture. In some cases, cells can be isolated that are not susceptible to the control processes involved in the regulation of cell proliferation, and these will continue to grow, pile up, and form *dense foci* of cells arranged in irregular crisscross patterns, on top of the sheet of normal cells. In most cases, such *in vitro* transformed cells are able to induce the formation of tumors when inoculated into susceptible animals and, therefore, have been thought to be comparable to tumor cells arising *in vivo*.

1. Isolation of Transformed Cells

Transformed cells have been reported to arise in several different systems *in vitro*. Malignant transformation was shown to occur readily after treatment of normal cells with oncogenic chemicals, radiation, RNA tumor viruses, and DNA tumor viruses. In most cases, the procedure used to isolate transformed cells depended upon the expression of one particular feature of these cells that is unshared by normal cells. It seems wise, therefore, to examine some of these properties in more detail.

2. Properties of Transformed Cells

The number of properties which are used to define the transformed state is regularly expanding, correlatively with our

TABLE II. External Cell Morphology [a]

Structures	Description	Location	Internal Structure	Appearance on Transformed Cells
Microvilli	Filamentous 0.1 μm diameter and up to 2 μm length	Mostly on top surface	Longitudinal, parallel 6-nm microfilaments	Varying numbers, lengths, distributions, and often distorted shapes
Filopodia	Filamentous 0.1 μm diameter up to 30 μm and longer	Often from margin, may be from top surface, extends to substrate and other cells	May contain micro filaments	Occur in unusual numbers and may be distorted
Ruffles	Sheetlike, 0.1 μm thick, up to 10 μm long, may be as wide as the cell	Mostly marginal at leading edge of migrating cell, occasionally on top surface	Microtrabecular lattice actin-rich	On margins and top surfaces, possibly several on a cell
Blebs (knobs)	Spherical and hemispherical, 0.2 μm to 5–10 μm in diameter	May appear any where on cell surface not adherent to substrate	Any cytoplasmic material	May appear in large numbers and persist for long duration, especially on unspread cells
Retraction fibers and attachment fibers	Filamentous, 0.1 μm diameter and wider, as long as 30 μm	From surface of spherical and highly convex cells, extends to substrate and other cells	May contain micro filaments	May be reduced in number, usually absent on poorly spread cells
Lobopodia	Peninsular strands of cytoplasm, cross section may be flat or round	From margin of flat cells or from spherical mass of mitotic cells	May contain filaments of cytoskeleton, such as stress fibers	Perhaps fewer on mitotic cells, absent on poorly spread cells

[a] Data from Allred and Porter (28).

knowledge in cellular biology. However, these transformation parameters may be pooled into three interrelated groups, according to whether they concern the morphology of the cells, growth control, or other molecular properties.

a. **Morphology of the Transformed Cell.** Many details concerning the morphology of transformed cells as compared with that of normal cells will be found in the excellent review by Allred and Porter (28). Briefly, transformed cells are characterized by a number of changes in criteria of the external morphology that are used to describe normal cells (Table II) and by a very high pleomorphism (i.e., numerous different shapes) in culture. In addition, the cell cytoskeleton organization (microtubules, intermediate filaments, and microfilaments) has been shown to be extensively altered in transformed cells. Disorganization and loss of cytoplasmic tubulin microtubules have been observed in SV40-transformed cells (29), and the organization of the microfilaments has been shown to be altered in most of the transformed cells in which these criteria have been studied (30–36) and in normal chick embryo fibroblasts (CEF) treated with concentrations as low as 10^{-10} M of the tumor promoter 12-0-tetradecanoyl phorbol-13-acetate (TPA) (37). However, Goldman et al. (38) did no detect any gross difference in the microfilaments' organization of SV40-transformed 3T3 (SV101) cells. In normal cells, microfilaments are composed of polymerized actin, myosin, tropomyosin-actinin, and filamin (or actin-binding protein) (30,31,39–55) organized in bundles (stress fibers, or cables) involved in cell motility (31,38,40,56–59) and cell-to-cell or cell-to-substratum interactions (30,31,34,60,61). Interestingly, the normal organization of actin cable is repaired in revertants isolated from SV40 transformants and in SV40–ts-mutant-transformed cells grown at the nonpermissive temperature (30,31,33,61–64).

b. **Cell Surface Organization and Composition.** A spectacular consequence of the cell surface changes, occurring upon cell transformation, is their ability to be preferentially agglutinated by lectins (65,66). These proteins, essentially isolated from plants, are able to combine specifically with the sugar moiety of glycolipids and glycoproteins of the cell surfaces and, hence, lead to the binding together of different cells. The increased susceptibility of transformed cells to lectin is not due to an elevated number of binding sites on the cell surface (67–69). The lectin receptors appear to be randomly distributed on both normal and transformed cells (70–76). It is thought that addition of lectin would rather aggregate the receptors in patches leading to the formation of caps at several sites on the cell surface. The redistribution of sites following the binding of lectins would be inhibited in normal cells (74,76–78). It appears unlikely that lipids play a controlling role in the cell agglutinability (79–83), whereas a mild treatment of normal cells by proteases can increase their agglutinability (84–86), suggesting that proteins are involved in this phenomenon.

c.   **Cell Surface Proteins in Transformed Cells.**   Many recent reviews have dealt with this rapidly expanding field of research, and we will try briefly to summarize the results obtained (for reviews, see refs. 87,88,89). The identification and comparison of the normal and transformed cell surface proteins after polyacrylamide gel electrophoresis has led to two types of proteins being distinguished: those whose synthesis is increased after transformation and those whose synthesis is decreased in transformed cells. Membrane proteins with molecular weights 90,000 to 95,000 and 73,000 to 79,000 daltons have been found to be present in increased amounts in transformed cells (90–92). A correlation between the accumulation of proteins with similar molecular weights (i.e., 75,000 and 95,000 daltons) and glucose starvation has been postulated (93,94). Furthermore, the synthesis of these proteins has been induced by inhibition of glycosylation (93), whereas feeding of transformed

TABLE III.   Loss of Fibronectin in DNA Tumor
Virus–Transformed Cells[a]

Transforming Virus	Cell Species	References
SV40	Human (WI 38), mouse (3T3)	112, 126, 128, 133–135
Polyoma	Hamster (NIL, BHK 21) mouse (embryo cells, 3T3, 3T6)	125–127, 133, 136
Adenovirus	Hamster (embryo kidney), rat (embryo fibroblasts, embryo myoblasts)	111
HSV	Hamster (NIL)	125, 137

[a] Data from Vaheri and Mosher (103).

cells with higher amounts of glucose may prevent the induction of their synthesis.

Many more proteins have been reported to be found in reduced amounts in transformed cells. Among them, the most studied are collagen and fibronectin.

Collagen is synthesized as precursor, procollagen, which is not entirely cleaved by cells (95). Both collagen and procollagen may be found incorporated in the extracellular matrix of the cells. The synthesis of collagen has been known for a long time to be depressed in transformed cells (96–100), probably because of a 5- to 10-fold decrease in the amount of procollagen mRNA synthesized, at least in chick fibroblasts (101). The properties and possible role of fibronectin have been extensively reviewed (see, for example, reviews by Yamada and Olden, ref. 102, and Vaheri and Mosher, ref. 103). The salient features of these properties are the following.

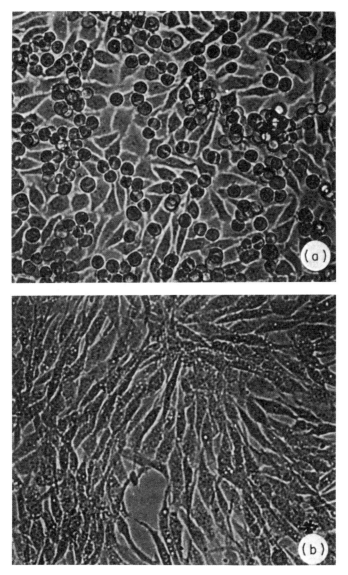

FIGURE 3. Transformed NIL8 HSV cells with and without LETS protein purified from conditioned medium of normal cells. *a*, Control cultures; note refractile nature of cells and presence of rounded detached cells; *b*, Plus 100 µg/ml LETS protein; note flattening, elongation, and alignment of cells and absence of floating cells.

197

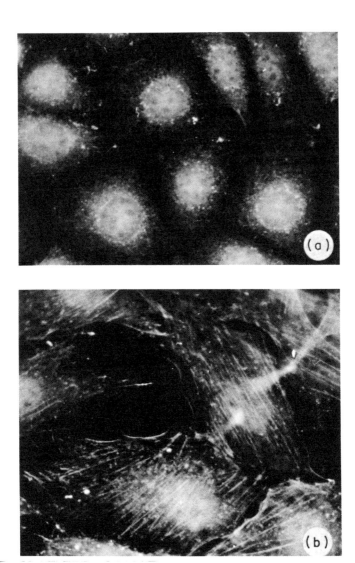

FIGURE 4. Effect of LETS protein on intracellular distribution of actin. Transformed NIL8 HSV cells were fixed and stained with antibody to actin. *a,* Control cells; note rounded polygonal shape of calls and absence of readily visible actin microfilament bundles; *b,* Cells treated with 50 µg/ml LETS protein; note the flattened asymmetric shape of the cells and the appearance of staining of extensive microfilament bundles.

198

Fibronectin (104,105) is a large external glycoprotein of molecular weight 210,000 to 270,000 daltons, also known as LETS protein (large, external, transformation-sensitive) (106) or cell surface protein (CSP) (107). It is located on the surface of the normal cell and arranged in fibrillar arrays (103,108–124). The external location of this protein has allowed its detection by surface radioactive labeling (125–132). Loss, or reduced amounts, of fibronectin have been found in most of the transformed cells from different origins (reviewed in 102–103; see also, Table III, page 196). The functions and the possible reasons for the absence of fibronectin in transformed cells have been discussed elsewhere (138). In cells transformed by thermosensitive viruses for the maintenance of transformation (see below), the reduction of cell-associated fibronectin takes place only at the permissive temperature (90,114,127,133–137,139,140). Furthermore, the tumorigenic potential of transformed cells has been found to correlate in several cases with reduced amounts of fibronectin (111,141,142). Several morphological aspects of the transformed cells can be reverted by the addition of purified fibronectin to the culture medium (114,122,143,146) (Figs. 3,4), whereas several other transformation parameters are still expressed in the presence of added fibronectin (114,122,144,145,147).

The reduced rate of fibronectin biosynthesis observed in transformed cells (148,149) has been shown to result from a reduced synthesis of the corresponding mRNA (101). An increased turnover of fibronectin in transformed cells has also been described (149).

The possible role of fibronectin and other matrix components in the process of malignant transformation has been extensively discussed in several reviews (see, for example, refs. 102,103,138).

d. Loss of Growth Controls. Transformed cells are not sensitive to the control processes that regulate the proliferation of normal cells and therefore exhibit properties that have been

valuable mostly for their isolation. Reduced serum requirements, loss of contact inhibition, and ability to grow without
anchorage have been used to select transformed cells.

i. REDUCED SERUM REQUIREMENTS. Whereas normal cells in
culture have generally high serum requirements (150–154),
transformed cells are able to grow at very reduced serum concentrations, sometimes in serum-free medium. This property
has been used by Smith et al. (155) to select SV40-transformed
BALB/c 3T3 cells. After infection with the virus, cells were

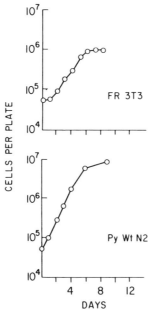

FIGURE 5. Saturation density of normal and transformed cells. 5×10^4
Cells of normal FR 3T3 rat cells and polyoma wild-type virus-transformed
rat cells have been used to seed the cultures. The medium was changed every
other day. The number of cells was counted every 2 days. Data from Seif
(14).

plated in a medium containing concentrations of serum that were not sufficient to allow the multiplication of normal parental cells. Only transformed cells would grow and therefore could be selected on this basis.

ii. ANARCHIC PROLIFERATION OF TRANSFORMED CELLS. Transformed cells do not seem to be susceptible to the contact inhibition of movement that characterizes normal cells (6,7) and they are able to form multilayers, leading to the appearance of dense foci on the monolayer of untransformed cells if a mixed culture is grown. As a consequence, transformed cell lines will grow at higher saturation densities (number of cells grown on a given surface) than will their untransformed counterparts (Fig. 5). This property will also lead to the formation of isolated dense colonies when the cultures are seeded with low cell densities, allowing transformed cell lines to be cloned. This technique has been widely used to isolate virus-induced transformants and offers the great advantage that it does not exert any particular selective pressure on the infected cells other than that exerted by growth on a plastic (or glass) substratum.

iii. ABILITY TO GROW WITHOUT ANCHORAGE. Normal cells do not multiply if they are grown in liquid or semisolid mediums. Based on the ability of established polyomavirus-transformed BHK-21 cells to grow in suspension in agar (156) is the selective procedure developed by McPherson and Montaginer (157) to isolate polyomavirus transformants from infected cells by cloning in agar. The procedure can be briefly outlined as follows. Virus-infected cells are resuspended, after infection, in a small volume of culture medium. An agar base layer (1.2% agar) containing cell culture medium and serum is poured in plastic petri dishes and left to set at room temperature. Infected cells are plated in agar (0.33%) or methocel (1.2%) on the top of the preceeding layer. After setting, plates are incubated in a moist atmosphere at the required temperature. Transformed cells will multiply and form large colonies that remain isolated because of the relative viscosity of the suspension, and untransformed cells will not divide. Colonies of transformed cells can be removed from the agar medium with Pasteur's pipettes and transferred in a tube containing a small volume of medium. After the agar has been broken into small parts by repeated pipetting, the cells are trypsinized and replated. This technique has been

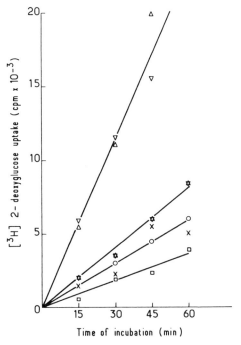

FIGURE 6. Kinetics of 2-deoxyglucose uptake in normal and transformed FR-3T3 cells. Cells were seeded on cover slips at a density of 5×10^4 cells and grown at 33°C. The cell slips were then incubated in the presence of 2-³H deoxyglucose. At the indicated times, the cover slips were quickly washed with isotonic buffer and the radioactivity associated with the cells was measured.   0, polyoma-transformed cells;   X, SV40-transformed cells; normal FR 3T3 rat cells. Data from Perbal and Rassoulzadegan (168).

used to isolate several transformants after infection by several tumor viruses.

e. Other Transformation Parameters. Reduced levels of cyclic AMP, enhanced rate of sugar transport, and increased protease activity are among the other characteristic properties of transformed cells.

i. CYCLIC NUCLEOTIDES. Changes in the cyclic nucleotides metabolism observed in transformed cells have been reviewed by Graham (158) and may be summarized as follows: low levels of cAMP have been found in many transformed cells (159,160),and adenyl cyclase activity has been found to be reduced in polyomavirus-transformed cells (159), whereas it was found to be increased in SV 3T3 cells (159,161) and to be similar to the normal value in polyomavirus 3T3–transformed cells (161).

ii. HEXOSE UPTAKE. Increase of hexose uptake has been reported to be associated with oncogenic transformation in several different systems (162–168) and to result from an accelerated synthesis of glucose carriers (169).

Enhanced sugar uptake may easily be detected by incubation of transformed cells in the presence of labeled 2-deoxyglucose or 3-0-methylglucose (Fig. 6).

iii. PROTEASES. The synthesis of many proteases has been observed to be increased after oncogenic transformation. One of them is a serine protease able to function, under the laboratory conditions generally used, as a plasminogen activator (PA).

The production of PA has been reported to correlate with malignancy in various cases (170–176), the appearance of PA activity being closely related with, and being an early event in, the transformation process (174,177–179). Increased PA activity has been detected in several transformation systems from different

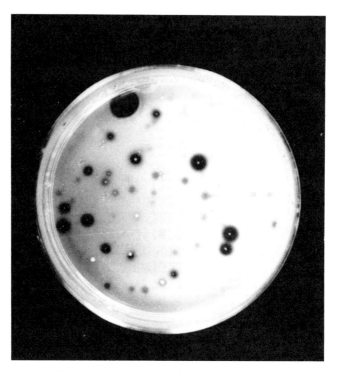

FIGURE 7.   Plasminogen activator induced careinolysis by polyoma-virus-transformed cells. Wild-type polomavirus-transformed cells were grown in soft agar. An overlay containing 1% agarose, 2.5% dried milk, and 20 μg of purified plasminogen was poured on the top of the layer containing the foci of transformed cells. After overnight incubation at 33°C, clear halos resulting from casein digestion were visible.

origins [see reviews by Reich (180), Christman, et al. (181), and Quigley (182)].

☐ *Detection of PA activity may be carried out either by qualitative or quantitative tests.*

Qualitative Tests: *Upon incubation in the presence of PA, plasminogen is activated in plasmin, a trypsinlike enzyme. The first assay, introduced by Goldberg (173), measures plasmin-mediated caseinolysis. A mixture containing agarose, nonfat dried milk, purified plasminogen, and serum is poured on the cells before incubation at 37°C. The hydrolysis of casein induced by plasmin will result in the formation of clear halos. This method has been used to detect PA production by polyomavirus and SV40-rat-transformed cells grown either in agar or plastic (168) (Fig. 7).*

The second qualitative assay, described by Jones et al. (183) detects plasmin-mediated fibrinolytic activity. The mixture poured on the cells contains agarose, serum, bovine thrombin, and fibrinogen. After overlay, the agarose solidifies and a transparent gel is formed which becomes cloudy because of the fibrin formation. Plasmin is able to induce lysis zones, which are easily detected on the background. Both of these tests may be performed with electrophoretically purified PA by pouring the mixture on top of acrylamide gels (189) and, therefore, allowing a comparison of PA produced by different cell lines.

Quantitative Tests: *Two different kinds of quantitative tests have been used, those which measure plasmin activity and those which measure directly the activation of plasminogen or the hydrolytic activity of PA.*

Plasmin activity has been quantitatively detected with the use of ^{125}I-labeled fibrin-coated plates (178) or ^{3}H-la-

beled casein (168) as substrates. Fluorometric assays and isotopic Tx ArgOMe (N-toluene sulfonyl-L-arginine methyl ester) assay have also been described (184). These methods cannot be used to determine the kinetic properties of PA, and other tests have been developed that measure PA activity directly. A direct fluorescent assay that uses a synthetic fluorogenic peptide substrate, 7(N-Cbz, glycylglcyl argininamido)-4 methylcoumain trifluoroacetate, has allowed the study of kinetic properties of PA produced by Rous sarcoma virus (RSV) transformed cells (185). Another approach consists of measuring the amount of plasmin generated in the presence of PA. Dano and Reich (186) used a reaction based on the conversion of ^{125}I-labeled plasminogen into ^{125}I-labeled plasmin. The substrate and product were separated by polyacrylamide gel electrophoresis and the radioactivity of each subsequently measured. This method allowed the study of inhibitors of the activation (plasminogen–plasmin) reaction. Another method is based on the different isoelectric points of heavy-chain (pH 4.9) and light-chain (pH 5.8, 5.9, and 6.0) human plasmin (187). ^{3}H-labeled purified human plasminogen incubated in the presence of PA gives rise to ^{3}H-labeled plasmin. After addition of 2-mercaptoethanol and iodoacetamide, the mixture of ^{3}H-labeled heavy and light chains is applied on a small column of SE-Sephadex A50 equilibrated with acetate buffer 0.1 M, pH 5.0. Under these conditions, light chains are eluted and may be directly counted after addition of Picofluor scintillation fluid. Heavy chains are eluted with borate buffer (pH 9.0). Unactivated PA is also eluted under these conditions. This method has been developed to study the levels of PA produced by polyomavirus (Py) and SV40 transformed rat cells under different growth conditions (B. Perbal, unpublished results).   □

PA production has been detected in SV40-transformed cells

(168,174,177,188–192,193), polyomavirus transformed cells, (168) and cytomegalovirus-transformed cells (194,195). PA production was shown to be an early event after infection of hamster or human cells with partially inactivated HSV$_2$ (196).

The level of PA produced by freshly isolated SV40 rat transformed cells (168) has been found to be of the same range as that of normal cells, and much lower than the level of PA produced by polyomavirus-transformed rat cells isolated and tested under the same conditions. Furthermore, the proportion of clones that produce PA was found to be very low in the case of SV40-transformed cells, whereas high percentages were observed among polyomavirus transformants (168). Further studies have shown that it is possible to isolate PA producer cells from SV40 PA nonproducer cell lines (B. Perbal, unpublished observations) by growing the nonproducer cells for a great number (100 to 150) of generations and that expression of PA production by polyomavirus-transformed cells may be counterselected when the infected cells are grown under certain conditions (197). Thus, the percentage of PA producers among polyomavirus transformants is about 10 times lower when the infected cells are kept in an actively growing state for several days after infection (197).

The synthesis of PA by transformed cells is predominantly determined by the cellular genome (198) and may be influenced by the expression of viral genes (168,178,197–199), while the levels of PA production by some normal cells was found to be subjected to physiological and developmental states of the cells (179,200–202) and modulated by several cyclic nucleotides, glucocorticoids, retinoids, lymphoid products, and growth factors (200,203–211). This may explain why some mammalian cells do not always produce increased levels of PA after transformation or why they differ in their response to different viruses (176,197,212–214).

f. **Tumorigenicity of Transformed Cells.** The ability of the

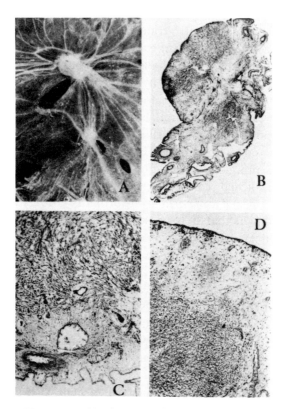

FIGURE 8. Tumors on Chorio Allantoid Membrane. *A,* RS2-TH2 cells. These cells are derived from a sarcoma induced by Schmidt-Ruppin–Rous sarcoma virus (SR-RSV) in hamster. Two confluent tumors are visible on top; one single tumor is located at the bottom, x 5; *B,* Same cells. The figure shows two nodules in mesoderm, × 25; *C,* RS2 cells. These hamster cells are derived from BHK-21/13 transformed with SR-RSV. The figure shows chorio allantoid membrane endoderm, ×125; *D,* L cells. These are mouse spontaneously transformed cells. The figure shows the chorio allantoid membrane ecotderm, ×125. Figures from *Int. J. Cancer* **4:** 859 (1969).

transformed cells to induce tumors (tumorigenicity) can be tested by one of the following methods.

(a) Infection of inbred strains of animals to reduce the problems due to the histocompatibility system which might cause the destruction of the infected cells. When inbred strains are not available, animals can be immunosuppressed either by irradiation or by injection of anti-lymphocytic serum.

(b) Infection of nude mice, being immunologically deficient because they are unable to produce T lymphocytes (involved in the rejection process). This assay has been shown to be a reliable one that distinguishes between normal and malignant cells (215).

(c) Inoculation of the chorioallantoic membrane of fowl eggs, which has been reported to be an immunologically depressed site within the first 2 weeks of incubation (216). This test allows a quantitative assay to be made of the tumorigenicity of transformed cells by counting tumors produced in 7 days in 11-day-old eggs (217) (Fig. 8).

Many attempts have been made to establish a correlation between the expression of one, or several, transformation parameters and the tumorigenic potential of transformed cells. Such studies are possible only when parental untransformed cells are available. Unfortunately, some of the established cell lines used in many different laboratories have induced tumors when tested under certain conditions. This is the case of 3T3 mouse cells and BHK-21 hamster cells (218,219). Different classes of SV40 transformants can be obtained after infection of primary embryo cells (33,174), established mouse 3T3 cells (174,220), and FR3T3 rat cells (168). The different phenotypes observed include various abilities to grow without anchorage (220), differ-

TABLE IV. Summary of the Results of Transformation Assays Performed on Various Derivatives from the FR 3T3 Cell Line[a,b]

Cell Line	Isolation[c]	Saturation Density[d]	Generation Time[e] (h)	Growth in Agarose Medium[f]	Growth in Low Serum[g]	Hexose Uptake Rate[h]	Membrane Protease Activity[i]	PA Production[j]
FR 3T3		5	25	–	–	Low	Low	–
SV-WT-N1	f	45	20	+	–	Medium	Low	–
SV-WT-N2	f	45	20	–	–	Medium	Low	–
SV-WT-A1	a	45	20	++	–	Medium	Low	–
SV-tsA30-N1	f	45	24	+	–	Medium	Low	–
SV-tsA30-N4	f	45	18	++	–	Medium	NT[k]	–
SV-tsA30-A1	a	45	15	++	–	Medium	Low	–
SV-tsA30-A2	a	45	19	++	–	Medium	NT	–
SV-tsA30-An3	f	45	16	+	–	NT	NT	NT
Py-WT-N2	f	45	20	+	+	High	High	+
Py-WT-N3	f	45	NT	+	+	High	High	+
Py-WT-A1	a	45	12	++	+	High	High	+
Py-tsa-N1	f	45	12	+	+	High	High	++
Py-tsa-N3	f	45	12	+	+	Medium	High	++
Py-tsa-N11	f	45	12	+	+	High	High	++
Py-tsa-A1	a	45	NT	++	+	High	High	+
Py-tsa-A2	a	45	12	++	+	High	High	+

[a] Reprinted from B. Perbal and M. Rassoulzadegan (168).

[b] Established from Fisher rat embryonic cells.

[c] a, Colonies in agarose; f, focus formation.

[d] Cell densities at the growth plateau. Cells $\times 10^4$ per square centimeter.

[e] Measured in Dulbecco-modified Eagle medium supplemented with 10% newborn calf serum.

[f] Measured by seeding 5×10^4 cells in soft agar and counting the colonies 8 days later. Growth ability in agarose was scored as follows: ++, more than 75% of the cell input gave rise to colonies; +, 25%; −, less than 0.002%.

[g] Symbols: −, slow growth and low saturation density similar to those observed in the presence of high serum concentrations; +, generation time and maximal cell density similar to those observed at high serum concentrations.

[h] Explanation of terms: low, values similar to that shown for FR 3T3 cells; medium, values in the same range as that shown for SV-WT-A1; high, values similar to that shown for Py-tsa-N1.

[i] Hydrolysis of [³H] casein in the absence of plasminogen after 8 hours of incubation at 33°C. Low, slight hydrolysis (very similar to that of FR 3T3); high, significantly higher rate of hydrolysis (can reach a value which is 10 to 15% that obtained in the presence of plasminogen).

[j] Symbols: ++, positive reaction by the agar overlay assay and rate of [³H]casein hydrolysis similar to that obtained for Py-tsa-N1 cells; +, rate of [³H]casein hydrolysis in the same range as that shown for Py-tsa-A1 cells; −, no detectable caseinolysis by either assay.

[k] NT, Not tested.

ent levels of deoxyglucose uptake (168,195), different levels of PA activity (195), and expression of viral tumor antigens (220).

Studies performed with the different types of transformants obtained after infection of primary rat embryo cells have revealed that minimal transformants were not tumorigenic when injected in nude mice, whereas fully transformed cells were highly tumorigenic (221–223). Distinct transformation phenotypes were also found to be induced after infection of FR 3T3 rat cells with polyomavirus and SV40 viruses. SV40 transformants exhibited intermediate levels of transformation, whereas polyomavirus transformants appeared to be fully transformed (Table IV). Interestingly, SV40-transformed cell lines were found to be much less tumorigenic than the polyomavirus transformants when tested under various conditions. It would seem, therefore, that tumorigenicity correlates with a fully transformed state.

Whether or not SV40 tumor cells are selected for the expression of a fully transformed phenotype, as suggested by previous studies (214), may be investigated by examination of the transformation phenotype exhibited by tumor cells derived from SV40 transformants from the different described classes. Two different polyomavirus FR 3T3 transformed cell lines were used to induce tumors on chicken chorioallantoic membranes, and the phenotype of tumor cells was compared with that of parental transformed cells. Rate of deoxyglucose uptake, PA activity, and cloning efficiency in agar were found to be similar to that of parental cells in both cases (B. Perbal, unpublished observations).

The isolation of PA producer cells from several SV40 nonproducer cell lines has allowed us to distinguish operationally three different classes of SV40 transformants: PA nonproducer (class I), PA producer when grown on plastic (class II), and PA producer when grown either on plastic or in agar (class III) (B. Perbal, unpublished observations). These different classes may correspond to discrete levels of transformation, and it would be of great interest to study the tumorigenic potential of these different transformants and to study the phenotype of tumor cells derived from these transformants.

Oncogenic transformation, leading to the formation of tumorigenic cells, has also been described with herpes simplex virus in many cases (224–234). Several classes of transformants have also been distinguished on the basis of their ability to induce tumors, and agar-selected transformed cells were found to be more tumorigenic than those isolated as dense foci (234). A study of tumor production by HSV_2-transformed rat cells inoculated into immunosuppressed rats (235) revealed that HSV_2-transformed cells were able to induce tumors with a latent period of nearly 2 years in these animals.

II. ESTABLISHMENT AND MAINTENANCE OF VIRAL-INDUCED TRANSFORMATION

A. HOST CELLS

One of the most important factors in the establishment of the transformed state is the genetic constitution of the infected cells. It has been shown, for example, that two classes of cells may be distinguished on the basis of their behavior upon infection with DNA tumor viruses. In permissive cells, all the viral functions necessary for the production of infectious progeny are expressed, and the lytic cycle, whose different characteristics have been developed in the preceding chapters, takes place. In nonpermissive cells, the multiplication of the infecting virus is blocked at an early stage, no infectious progeny are produced, and the cell may be stably transformed. The molecular basis for cell permissivity is actually unknown. However, in many cases, the situation is not so strictly defined, and semipermissive cells very often have been described. In such cultures, some cells are permissive and support the lytic viral multiplication and some others are nonpermissive. Among nonpermissive or semipermissive cells for the multiplication of SV40 are rat cells (236), bovine cells (237), and guinea pig cells (236). African green monkey cells are permissive for polyomavirus and appear to be semipermissive for SV40. Rat and hamster cells are semipermissive for polyomavirus (reviewed in 238).

Several herpesviruses have been found to grow and induce cytopathic effects in almost all cell lines tested. Therefore, many different manipulations have been used to abolish the virus infectivity and isolate herpes-transformed cells. These include: UV irradiation of the virus (224,225,230,239,240), photodynamic inactivation of virus treated by heterocyclic dyes (241), use of temperature-sensitive mutants at the nonpermissive temperature (227,229,242), and infection at high supraoptimal temperatures (232,243,244). On the other hand, Esptein-Barr virus (EBV) exhibits a very restricted host range specificity, and no cell line has been found yet to be fully permissive to EBV. Long-term lymphoblastoid cell lines have been obtained after infection of B lymphocytes with EBV, either *in vivo* or *in vitro* (245–251). The host range of EBV *in vitro* is restricted to primate B lymphocytes (252–253) which seem to be semipermissive for the multiplication of EBV as a small proportion of the infected cells support viral replication. This state has also been called "restringent infection" (236). Finally, marmoset lymphocytes appear to be more permissive for viral replication than do human lymphocytes (255,256). EBV strain HR-1, obtained from the HR-1 clone (257), or a culture established from a Burkitt's tumor biopsy (jijoye cells) lack the ability to transform (256,258), whereas strain B95-8 obtained from a clone of marmoset cells (originally infected with EBV from a patient with infectious mononucleosis, ref. 255) are able to induce transformation. This situation may result either from a virus defect (see below) or from a cellular restriction, or from both.

In vitro cell transformation has also been reported to occur with cytomegalovirus (259–261), guinea pig virus (262), and equine herpesviruses (240). The oncogenic properties of Marek's disease virus, *H. saimiri, H. ateles,* Lucké herpesvirus, Epstein-Barr virus, and cytomegalovirus *in vivo* have been reviewed by Zur Hausen (263).

B. MULTIPLICITY OF INFECTION AND PHYSIOLOGICAL STATE OF THE CELLS

It is generally assumed that a multiplicity of infection (MOI) varying from 100 to 5000 pfu/cell will allow the isolation of DNA tumor-virus-transformed cells. The number of transformants obtained appears to vary linearly with the number of infectious particles when low multiplicities of infection are used (157,219,264). Transformation frequencies observed after infection of FR 3T3 rat cells at a MOI of 100 pfu/cell with several different strains of *Polyomavirus* were found to range between 0.12 to 0.15% (14,197,264,266), whereas the transformation frequencies obtained for an MOI varying from 30 to 100 pfu/cell with SV40 under similar conditions were found to range between 0.4 and 0.8% (267).

The use of thermosensitive mutants altered in the early (A/a) region of polyomavirus (Py) and SV40 genomes has established that both multiplicity of infection and physiological state of the infected cells may be important in the establishment of the transformed phenotype. Infections performed at the nonpermissive temperature have revealed that the early regions of SV40 and Py are involved in the establishment of transformation (268–272). Under these conditions, the frequency of transformants obtained with ts mutants is drastically reduced as compared to that obtained with the corresponding wild type (wt) strains.

The early gene products of Py and SV40 have also been implicated in the maintenance of the transformed state (14,197,267,273–278). The use of different multiplicities of infection showed that high MOI will favor the isolation of transformants in which some viral-induced growth controls are not operating and, therefore, that will grow in agar at both restrictive and permissive temperatures (279). Similar results were obtained with SV40 and polyomavirus rat transformants isolated from FR-3T3-infected cells kept in growth-inhibiting conditions (confluency or growth in agar) (14,197,267,278). On the contrary, when the infected cells were seeded at low densities in culture and were kept in an actively growing state for several days after infection, most of the resulting isolated transformants were unable to grow in agar at the restrictive temperature (14,197,267,278,279).

PA producton has been studied in different types of transformants. The proportion of PA producer cells among transformants isolated in both conditions has been found to be dramatically reduced when the infected cells are kept in an actively growing state after infection, compared with the proportion obtained when the infected cells are incubated in growth-inhibiting conditions (confluency or agar). With SV40 and other polyomaviruses, a 10-fold decrease in the proportion of PA producers has been observed (197). These observations suggest that different populations of cells, which may not even be comparable, are selected by the different growth conditions applied to the infected cells. Furthermore, it has been shown that PA production may be under the control of a viral (probably large T antigen) in transformed cells whose ability to grow in agar is not under such a control (195,197), suggesting that ability to grow in agar and PA production are independently regulated in transformed cells (see below).

III.   EXPRESSION OF VIRAL PRODUCTS IN TRANSFORMED CELLS

A.   STATE OF THE VIRAL DNA TUMOR IN HERPESVIRUS-TRANSFORMED CELLS

Some EBV DNA has been found to be integrated into the DNA of restringently infected cells (280,281). However, these cells contain many copies of EBV DNA [representing more than 90% of the viral genome (282–284), in a circular form (285–287]. It has been known for some time that EBV B95-8 strain is able to induce transformation, whereas strain HR-1 is unable to transform (see above). Interestingly, the HR-1 genome has been reported to miss a portion of DNA (HSV IB fragment) which is present in B95-8 strain (288) and therefore would be directly involved in transformation. Both integrated and free virus DNA has also been described in Marek's-disease-transformed cells grown from an ovarian lymphoma and from a splenic lymphoma of chicken with Marek's disease (289), and circular DNA has been found in a circular structure in *H. saimiri*–transformed cells (290).

In the case of herpes simplex virus-transformed cells, the situation is complicated by the fact that transformed cells have sometimes been obtained with fragmented DNA or have been selected for the expression of a particular function. This is the case of TK$^+$ transformants isolated after infection of TK-negative cells by UV-irradiated virus (291–293) or after transfection of the TK$^-$ cells with sheared DNA or purified restriction endonuclease fragments (294–297). Such transformants have been found to contain DNA fragments which may account for 3 to 22% of the HSV genome. One to five fragments of HSV DNA per cell have been detected in these cells, indicating that a single fragment is sufficient to induce biochemical transformation.

The TK synthesized in these transformed cells is immunologically and biochemically identical to the viral-induced thymidine kinase detected during the lytic cycle (293,298–301).

Analysis of the HSV DNA content of several TK$^+$ transformants, obtained by either method, shows that HSV DNA sequences are common to all the transformed cells tested. These sequences map between 0.28 and 0.32 units (301), whereas DNA sequences arising from the regions between 0.11 to 0.57 and 0.82 to 1.00 map units were also randomly integrated in the host DNA. It appears that fragments located between 0 to 0.11 and 0.57 to 0.52 map units have never been found in the transformants tested (Fig. 9). These results suggest that a strong negative selective pressure is counterselecting the integration of these sequences, whose expression therefore might be involved in the lytic cycle and interfere with the establishment or the maintenance of the transformed state.

The study of revertant TK$^-$ clones selected either after exposure to ^{3}H-labeled thymidine or for resistance to Budr has revealed that in most cases the TK$^-$ phenotype is accompanied by the loss of viral DNA (302). However, in one case, the TK$^-$ line was still found to contain a single copy of the viral DNA fragment (302). The results of somatic cell genetic analysis performed with a transformed mouse cell line suggest that the viral TK gene is integrated into a murine chromosome at a site which seems different at the region coding for murine TK (303).

HSV-transformed cells have also been obtained on the basis of their growth properties. This is the case for a HSV$_2$ transformed hamster cell line which was originally isolated as dense foci after infection of the parental cells by UV-irradiated virus (304). Twenty independent clones of this transformed cell line have been isolated (305) and tested for their HSV-DNA content (306). The results obtained show that, in all cases, a portion of the viral DNA is integrated into the host cell DNA and that a few copies of the viral sequences are present per cell. Two sets of sequences located between 0.21 to 0.33 and 0.60 to 0.65 map units on the HSV_2 genome have been constantly found in all the clones tested (Fig. 9).

The use of the HSV_1 and HSV_2 restriction endonuclease fragments has allowed the definition of two regions of the genome exhibiting transforming ability. One region has been located at map position 0.30 to 0.45 on the HSV genome (307), and another region has been located at map position 0.58 to 0.62 on the HSV_2 genome (308). These morphological transforming regions (mtr I and mtr II) have no homology, and the 15.8 kb mtr I region that maps in the unique long region of HSV genome is responsible for morphological transformation of mammalian

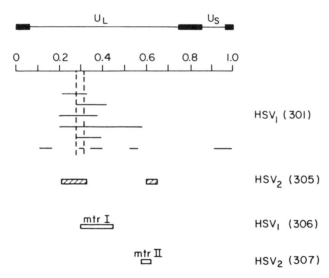

FIGURE 9.   Physical map for HSV DNA in transformed cells. The DNA fragments found in HSV_1 TK⁺ transformed cells (301) and DNA sequences found in all HSV_2 transformed cells (305) are reported. In addition, the location of morphological transforming regions of HSV_1 and HSV_2 are indicated (mtr I and mtr II).

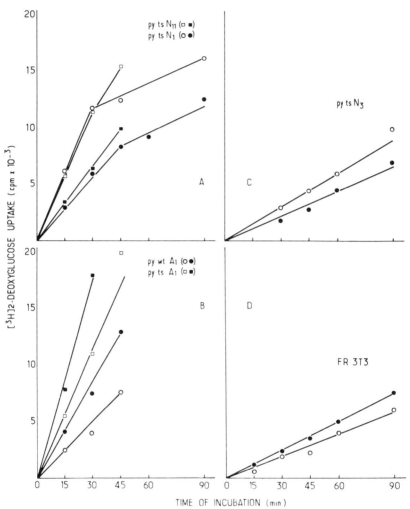

FIGURE 10. Thermosensitivity of 2-deoxyglucose uptake in polyoma-transformed cells. The rates of deoxyglucose uptake at 38°C (open symbols) or 41°C (closed symbols) have been determined as described in Figure 5. Data from Perbal and Rassoulzadegan (168).

cells *in vitro* (308). Physical and genetic mapping have shown that eight different functions are encoded by this region.

A first group of mutants mapping at 0.315 to 0.422 show a block at a level of transcriptional function required for viral DNA synthesis. These mutants are DNA (-) and have a normal DNA polymerase activity. They belong to complementation group 1-1.

The second group of mutants map at position 0.386 to 0.418. These mutants, which belong to complementation groups 1-3 and 1-4, have a thermolabile viral DNA polymerase activity and therefore exhibit a DNA(-) phenotype.

The third group (represented by mutants tsB5 and tsJ12) are affected in the structural genes for glycoproteins gA and gB.

Five other complementation groups also map in the mtr I region. Among them are two mutants with a DNA (+) phenotype (tsA, txF strain 17) and three DNA-deficient mutants (txM19, tx022, tsP23 strain KOS).

B. VIRAL RNA IN TRANSFORMED CELLS

Specific HSV RNA sequences have been detected in HSV_2 transformed hamster cells isolated as foci (309) and in TK^+HSV_1 biochemically transformed mouse cells which also exhibit a morphologically transformed phenotype (293). In both cases, cells had been infected with UV-irradiated virus. The results obtained indicate that a very small fraction (10–13%) of the viral sequences are transcribed in HSV-transformed cells, and that viral sequences are found mainly as cytoplasmic polyadenylated RNA (see Table V). HSV-specific RNA species have also been found in human cervical squamous cell neoplasms (R.P. Eglin, International Workshop on Herpes Viruses, July 27–31, 1981, Bologna, Italy).

The viral RNA content of Burkitt's tumor tissue ,has been

studied. The results obtained show that the tumor tissue contains RNA homologous to 3 to 6% of EBV DNA and that most of these RNA species are polyadenylated (310). Of particular interest is the observation that HR1 Burkitt's tumor isolates specify an abundant RNA class which is coded by the additional DNA found in W91 and HR1 strains. The viral RNA content of two cell lines (raji and namalwa) established by growing Burkitt's tumor biopsy cells *in vitro* has also been investigated. The polyribosomal and poly-A + RNA from these cells contain RNA complementary to 5 to 7% of EBV DNA (254,311), and the location of the DNA fragments coding for these DNA species has been determined (312). The most abundant RNA in namalwa cells has been found to be encoded by a DNA fragment whose coding capacity would correspond to the EBV nuclear antigen (EBNA) detected in these cells (313,314).

C. VIRAL PROTEINS IN TRANSFORMED CELLS

Several viral-specific antigens have been described in herpesvirus-transformed cells (see reviews by Roizman and Kieff, ref. 320, and Roizman and Spear, ref. 315). These antigens are characterized far less than the tumor antigens induced by papovaviruses and adenoviruses, and more studies are needed to determine whether they are involved in the establishment or maintenance of transformation.

D. VIRAL PRODUCTS NEEDED FOR THE MAINTENANCE OF TRANSFORMATION

The possible role of the different early and late genes in the maintenance of papovavirus-induced transformation has been investigated by the use of different mutant viruses whose defects were found to be located in different part of the genome.

TABLE V. Hybridization of Total Labeled RNA from Clone "i" and from LMTK⁻ Cells to HSV_1 and Phage TI DNAs

Source of ³H-RNA		ct/min Bound to Filters				% ct/min Input Hybridized to HSV_1 DNA	
		TI DNA (5 g)		HSV_1 DNA (5 g)			
Clone "i" Nuclear RNA	poly-A(−)	177	282[a]	730	807[a]	0.0075	0.008[a]
		195	289[a]	784	745[a]		
		306	100[a]	987	1128[a]		
	poly-A(+)	71	56[a]	289	240[a]	0.159	0.169[a]
		53	44[a]	289	282[a]		
		67	—	284	—		
Clone "i" Cytoplasmic RNA	poly-A(−)	470	576[a]	1254	1409[a]	0.017	0.016[a]
		606	514[a]	1692	1885[a]		
		749	—	1805	—		
	poly-A(+)	77	60[a]	196	154[a]	0.161	0.152[a]
		63	67[a]	178	139[a]		
		53	50[a]	150	175[a]		
LMTK⁻ Total RNA	poly-A(−)	831	1943[a]	660	1050[a]	0.0072	
		1183	1581[a]	1543	1259[a]		
	poly-A(+)	80	62[a]	98	105[a]	0.0045	
		59	—	99	—		

[a] Indicates counts hybridized after samples had been heated at 80°C for 5 min in 60% (v/v) formamide, 2 × SSC.

Several reports have led to the conclusion that the late genes coding for the viral proteins VP1, VP2, and VP3 are not involved either in establishment or maintenance of the transformed state (269,270,272,277,316), and the early region was, therefore, thought to be required for transformation.

Cells of different origins, including primary embryo cells and established cell lines, have been transformed after infection by SV40 ts mutants altered in the early (A) region (tsA) (see above). These transformants exhibit several distinct parameters

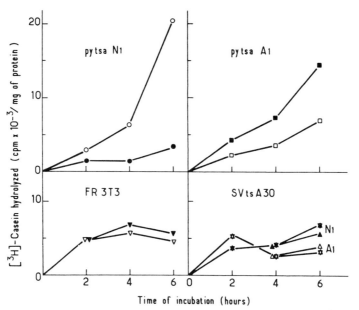

FIGURE 11. Plasminogen activator synthesis by Py-and SV40-transformed cells at either 33°C or 41°C. The production of plasminogen activator has been measured by a quantitative test based upon hydrolysis of ³H casein (see text). Polyomavirus-transformed cell lines exhibiting a thermosensitive phenotype (Py-tsa-N1) or a temperature-independent phenotype (Py-tsa-A₁) were compared to SV40-transformed cells of both types. PA production by normal cells at both temperatures is given by FR 3T3 cell lines. Data from Perbal and Rassoulzadegan (168).

of transformation in a temperature-sensitive way. These parameters include morphology (273,276,277), saturation density (168,273,276), growth in low serum (275), ability to grow in agar (168,273,274), presence of T antigen (273,276,277), increased uptake of sugar (168,273), and production of plasminogen activator (197) (Fig. 10, page 220 and 11, page 224). These observations indicate that active A gene product (or products) is (or are) required for the maintenance of transformation. The conflicting results that have been reported for the need of active early genes product in the maintenance of polyomavirus transformation might be due to the selection of different types of transformant (due to the different isolation procedures used) from a common bulk of infected cells. Thus, the growth state of the infected cells appears as a very powerful tool to select (or counterselect) a particular type of transformant (197).

The expression of several parameters of transformation has been reported to be under the control of large T antigen in polyomavirus-transformed cells (168). However, it would seem that such a situation is fortuitous, as shown by the fact that deoxygluxose uptake, PA production, and ability to grow in agar have been shown to be independently regulated in SV40 and polyomavirus-transformed cells (197). Finally, Seif and Martin (317) reported that SV40 small t antigen is not required for the maintenance of transformation but, rather, would act as a co-carcinogen (or promoter) during the process of transformaton establishment. Lania et al. (318) have studied the expression of several transformation parameters in polyomavirus-transformed cells and have concluded that large T antigen is not required for the maintenance of transformation.

REFERENCES

1. Rawls, W.E. and Adam, E. (1973), in *The Herpes viruses* (A.S. Kaplan, Ed.), Academic Press, New York and London.

2. Pimentel, E. (1979), *Biochim. Biophys. Acta* **560**:169.

3. Fleckenstein, B. (1979), *Biochim. Biophys. Acta* **560**:301.

4. Nazerian, K. (1979), *Biochim. Biophys. Acta* **560**:375.

5. Kruse, P.F. and Patterson, M.K., Jr. (Eds.) (1973), *Tissue Culture: Methods and Applications,* Academic Press, New York.

6. Abercrombie, M. and Heaysman, J.E.M. (1954), *Exp. Cell Res.* **6**:293.

7. Abercrombie, M. (1970) *In Vitro* **6**:128.

8. Haff, R.F. and Swim, H.E. (1956), *Proc. Soc. Exp. Biol. Med.* **93**:200.

9. Hayflick, L. and Moorhead, P.S. (1961), *Exp. Cell Res.* **25**:585.

10. Todaro, G.J. and Green, M.D. (1963), *J. Cell Biol.* **17**:290.

11. Foley, G.E., Drolet, B.P., McCarthy, R.E., Goulet, K.A., Dokos, J.M., and Filler, D.A. (1960), *Caner Res.* **20**:930.

12. Clausen, J.J., Syverton, J.T. (1962), *J. Natl. Cancer Inst.* **28**:1.

13. Puck, T.T., Ciecivra, S.J., and Robinson, A. (1958), *J. Exp. Med.* **108**:945.

14. Seif, R. (1977), These de doctorat de 3ᵉ cycle. Université de Nice.

15. Ford, C.E., Micklem, H.S., and Gray, S.M. (1959), *Br. J. Radiol.* **32**:280.

16. Krohn, P.L. (1962), *Proc. Roy. Soc. B.* **157**:128.

17. Cudkowicz, G., Upton, A.C., and Shearer, G.M. (1964), *Nature* **201**:165.

18. Siminovitch, L., Till, J.E., and McCulloch, E.A. (1964), *J. Cell. Comp. Physiol.* **64**:23.

19. Daniel, C.W., deOme, K.B., and Young, J.T. (1968), *Proc. Natl. Acad. Sci. USA* **61**:53.

20. Daniel, C.W., Young, L.J.T., and Medina, D. (1971), *Exp. Gerontol.* **6**:95.

21. Levan, A. and Biesele, J.J. (1958), *Ann. N.Y. Acad. Sci.* **71**:1022.

22. Rothfels, K. and Parker, R.C. (1959), *J. Exp. Zool.* **142**:507.

23. Hsu, T.C., Billen, D., and Levan, A. (1961), *J. Natl. Cancer Inst.* **27**:515.

24. Ponten, J. (1976), *Biochim. Biophys. Acta* **458**:397.

25. Yerganian, G. and Leonard, M.J. (1961), *Science* **133**:1600.

26. Ruddle, F.H. (1961), *Cancer Res.* **21**:885.

27. Macpherson, I. and Stoker, M. (1962), *Virology* **16**:147.

28. Allred, L.E. and Porter, K.R. (1979), in *Surfaces of Normal and Malignant Cells* (R.O. Hynes, Ed.), John Wiley & Sons, New York.

29. Wiche, G., Furtner, R., Steinhaus, N., and Cole, R.D. (1979), *J. Virol.* **32**:47.

30. McNutt, N.S., Culp, L.A., and Black, P.H. (1971), *J. Cell Biol.* **50**:691.

31. McNutt, N.S., Culp, L.A., and Black, P.H. (1973), *J. Cell Biol.* **56**:412.

32. Weber, K., Lazarides, E., Goldman, R.D., Vogel, A., and Pollack, R. (1974), *Cold Spring Harbor Symp. Quant. Biol.* **39**:363.

33. Pollack, R., Osborn, M., and Weber, K. (1975), *Proc. Natl. Acad. Sci. USA* **72**:994.

34. Bowman, P.D. and Daniel, C.W. (1975), *Mech. Ageing Dev.* **4**:147.

35. Edelman, G.M. and Yahara, I. (1976), *Proc. Natl. Acad. Sci. USA* **73**:2047.

36. Tucker, R.W., Sanford, K.K., and Frankel, F.R. (1978), *Cell* **13**:629.

37. Rifkin, D.B., Crowe, R.M., and Pollack, R. (1979), *Cell* **18**:361

38. Goldman, R.D., Yerna, M.J., and Schloss, J.A. (1976), *J. Supramol. Struct.* **5**:155.

39. Ishikawa, H., Bischoff, R., and Holtzer, H. (1969), *J. Cell Biol.* **43**:312.

40. Wessells, N.K., Speoner, B.S., Ash, J.F., Bradley, M.O., Ludvena, M.A., Taylor, E.L., Wrenn, J.T., and Yamada, K.M. (1971), *Science* **171**:135.

41. Perdue, J.F. (1973), *J. Cell Biol.* **58**:265.

42. Goldman, R.D. and Knipe, D.M. (1972), *Cold Spring Harbor Symp. Quant. Biol.* **37**:523.

43. Pollard, T.D. and Weihing, R.R. (1974), *CRC Crit. Rev. Biochem.* **2**:1.

44. Lazarides, E. and Weber, K. (1974), *Proc. Natl. Acad. Sci. USA* **71**:2268.

45. Weber, K. and Groeschel-Stewart, V. (1974), *Proc. Natl. Acad. Sci. USA* **71**:4561.

46. Hartwig, J.H. and Stossel, T.P. (1975), *J. Biol. Chem.* **250**:5696.

47. Goldman, R.D., Lazarides, E., Pollack, R., and Weber, K. (1975), *Exp. Cell Res.* **90**:333.

48. Lazarides, E. and Burridge, K. (1975), *Cell* **6**:289.

49. Lazarides, E. (1975), *J. Histochem. Cytochem.* **23**:507.

50. Mooseker, M.S. and Tilney, L.G. (1975), *J. Cell Biol.* **67**:725.

51. Sanger, J.W. (1975), *Proc. Natl. Acad. Sci. USA* **72**:1913.

52. Sanger, J.W. (1975), *Proc. Natl. Acad. Sci. USA* **72**:2451.

53. Wang, K., Ash, J.F., and Singer, S.J. (1975), *Proc. Natl. Acad. Sci. USA* **72**:4483.

54. Ash, J.F. and Singer, S.J. (1976), *Proc. Natl. Acad. Sci. USA* **73**:4575.

55. Heggeness, M.H., Want, K., and Singer, S.J. (1977), *Proc. Natl. Acad. Sci. USA* **74**:3883.

56. Goldman, R.D. and Follett, A.C. (1969), *Exp. Cell Res.* **57**:263.

57. Goldman, R.D., Pollack, R., and Hopkins, N.H. (1973), *Proc. Natl. Acad. Sci.* **70**:750.

58. Heaysman, J.E.M. and Pegrum, S.M. (1973), *Exp. Cell Res.* **78**:71.

59. Albrecht-Buehler, G. (1977), *J. Cell Biol.* **72**:595.

60. Abercrombie, M., Heaysman, J.E.M., and Pegrun, S.M. (1971), *Exp. Cell Res.* **67**:359.

61. Vollet, J.J., Brugge, J.S., Noonan, C.H., and Butel, J.S. (1977), *Exp. Cell Res.* **105**:119.

62. Osborn, M. and Weber, K. (1975), *J. Virol.* **15**:636.

63. Altenburg, B.C., Somers, K., and Steiner, S. (1976), *Cancer Res.* **36**:251.

64. Steinberg, B., Pollack, R., Topp, W., and Botchan, M. (1978), *Cell* **13**:19.

65. Burger, M.M. (1973), *Fed. Proc.* **32**:91.

66. Nicolson, G.L. (1974), *Int. Rev. Cytol.* **39**:89.

67. Cline, M.J. and Livingston, D.C. (1971), *Nature* **232**:155.

68. Ozanne, B. and Sambrook, J. (1971), *Nature* **232**:156.

69. Nicolson, G.L., Lacorbiere, M., and Eckhart, W. (1975), *Biochemistry* **14**:172.

70. Sela, B.A., Lis, H., Sharon, N., and Sachs, L. (1971), *Biochim. Biophys. Acta* **249**:564.

71. Arndt, Jovin, D.J., and Berg, P. (1971), *J. Virol.* **8**:716.

72. Martinez-Palomo, A., Wickner, R., and Bernhard, W. (1972), *Intl. J. Cancer* **9**:676.

73. Nicolson, G.L. (1973), *J. Natl. Cancer Inst.* **50**:1443.

74. Rosenblith, J.Z., Ukenq, T.E., Yin, H.H., Berlin, R.D., and Karnovsky, M.J. (1973), *Proc. Natl. Acad. Sci. USA* **70**:1625.

75. De Petris, S., Raff, M.C., and Mallucci, L. (1973), *Nature* **244**:275.

76. Ukena, T.E., Borysenko, J.Z., Karnovsky, M.J., and Berlin, R.D. (1974), *J. Cell Biol.* **61**:70.

77. Nicolson, G.L. (1973), *Nature* **243**:218.

78. Ash. J.F., Vogt, P.K., and Singer, S.J. (1976), *Proc. Natl. Acad. Sci. USA* **73**:3603.

79. Quigley, J.P., Rifkin, D.B., and Reich, E. (1971), *Virology* **46**:106.

80. Quigley, J.P., Rifkin, D.B., and Reich, E. (1972), *Virology* **50**:550.

81. Yau, T.M. and Weber, M.J. (1972), *Biochem. Biophys. Res. Commun.* **49**:114.

82. Gaffney, B.J., Branton, P.E., Vickus, G.G., and Hirschberg, C.B. (1976), in *Viral Transformation and Endogenous Viruses* (A.S. Kaplan, Ed.), p. 97, Academic Press, New York.

83. Gaffney, B.J. (1975), *Proc. Natl. Acad. Sci. USA* **72**:664.

84. Burger, M.M. (1969), *Proc. Natl. Acad. Sci. USA* **62**:994.

85. Inbar, M. and Sachs, L. (1969), *Proc. Natl. Acad. Sci. USA* **63**:1418.

86. Nicolson, G.L. and Blaustein, J. (1972), *Biochim. Biophys. Acta* **266**:543.

87. Burger, M.M. (1971), in *Current Topics in Cellular Regulation* (B.L. Horecker and E.R. Stadtman, Eds.), Vol. 4, p. 135, Academic Press, New York.

88. Hynes, R.O. (1976), *Biochim. Biophys. Acta* **458**:73.

89. Hynes, R.O. (Ed.) (1979), *Surfaces of Normal and Malignant Cells,* John Wiley & Sons, New York.

90. Stone, K.R., Smith, R.E., and Jolik, W.K. (1974), *Virology* **58**:86.

91. Isaka, T., Toshida, M., Owada, M., and Toyoshima, K. (1975), *Virology* **65**:226.

92. Chen, Y.C, Hayman, M.J., and Vogt, P.K. (1977), *Cell* **11**:513.

93. Pouyssegur, J., Shiu, R.P.C., and Pastan, I. (1977), *Cell* **11**:941.

94. Shiu, R.P.C., Pouyssegur, J., and Pastan, I. (1977), *Proc. Natl. Acad. Sci. USA* **74**:3840.

95. Bornstein, P. (1974), *Ann. Rev. Biochem.* **43**:567.

96. Green, H., Todaro, G.J., and Goldberg, B. (1966) *Nature* **209**:916.

97. Tsai, R.L. and Green, H. (1972), *Nature* **237**:171.

98. Peterkofsky, B. and Prather, W.B. (1974), *Cell* **3**:291.

99. Levinson, W., Bhatnagar, R.S., and Liu, T.Z. (1975), *J. Natl. Cancer Inst.* **55**:807.

100. Kamine, J. and Rubin, H. (1977), *J. Cell Physiol.* **92**:1.

101. Adams, S.L., Sobel, M.E., Howard, B.H., Olden, K., Yamada, K.M., DeCombrugghe, B., and Pastan, I. (1977), *Proc. Natl. Acad. Sci. USA* **74**:3399.

102. Yamada, K.M. and Olden, K. (1978), *Nature* **275**:179.

103. Vaheri, A. and Mosher, D. (1978), *Biochim. Biophys. Acta* **516**:1.

104. Keski-Oja, J., Mosher, D.F., and Vaheri, A. (1976), *Cell* **9**:29.

105. Engvall, E. and Ruoslahti, E. (1977), *Inst. J. Cancer* **20**:1.

106. Hynes, R.O. and Bye, J.M. (1974), *Cell* **3**:113.

107. Yamada, K.M. and Weston, J.A. (1974), *Proc. Natl. Acad. Sci. USA* **71**:3492

108. Ruoslahti, E., Vaheri, A., Kuusela, P., and Linder, E. (1973), *Biochim. Biophys. Acta* **322**:352.

109. Ruoslahti, E., Vaheri, A. (1974), *Nature* **248**:789.

110. Wartiovaara, J., Linder, E., Ruoslahti, E., and Vaheri, A. (1974), *J. Exp. Med.* **140**:1522.

111. Chen, L.B., Gallimore, P.H., and McDougall, J.C. (1976), *Proc. Natl. Acad. Sci. USA* **73**:3570.

112. Vaheri, A., Ruoslahti, E., Westermark, B., and Ponten, J. (1976), *J. Exp. Med.* **143**:64.

113. Yamada, K.M., and Pastan, I. (1976), *Trends Biochem. Sci.* **1**:222.

114. Ali, I.U., Mautner, V.M., Lanza, R.P., and Hynes, R.O. (1977), *Cell* **11**:115.

115. Bornstein, P. and Ash, J.F. (1977), *Proc. Natl. Acad. Sci. USA* **74**:2480.

116. Chen. L.B., Maitland, N., Gallimore, P.H., and McDougall J.C. (1977), *J. Exp. Cell Res.* **106**:39.

117. Mautner, V.M. and Hynes, R.O. (1977), *J. Cell Biol.* **75**:143.

118. Stenman, S., Wartiovaara, J., and Vaheri, A. (1977), *J. Cell Biol.* **74**:453.

119. Yamada, K.M., Yamada, S.S., and Pastan, I. (1977), *J. Cell Biol.* **74**:649.

120. Birdwell, C.R., Gospodarowicz, D., and Nicolson, G.L. (1978), *Proc. Natl. Acad. Sci. USA* **75**:3273.

121. Bornstein, P., Duksin, D., Balian, G., Davidson, J., and Crouch, E. (1978), *Ann. N.Y. Acad. Sci.* **312**:93.

122. Chen, L.B., Murray, A., Segal, R.A., Bushnell, A., and Walsh, M.L. (1978), *Cell* **14**:377.

123. Jaffe, E.A. and Mosher, D.F. (1978), *J. Exp. Med.* **147**:1779.

124. Hedman, K., Vaheri, A., and Wartiovaara, J. (1978), *J. Cell Biol.* **76**:748.

125. Hynes, R.O. (1973), *Proc. Natl. Acad. Sci. USA* **70**:3170.

126. Hoqq, N.M. (1974), *Proc. Natl. Acad. Sci. USA* **71**:489.

127. Gahmberg, C.G., Kiehn, D., and Hakomori, S.-I. (1974), *Nature* **248**:413.

128. Critchley, D.R. (1974), *Cell* **3**:121.

129. Gahmberg, C.G. and Hakomori, S.-I. (1975), *J. Biol. Chem.* **250**:2447.

130. Jone, C. and Hager, L.P. (1976), *Biochem. Biophys. Res. Commun.* **68**:16.

131. Critchley, D.R., Wyke, J.A., and Hynes, R.O. (1976), *Biochim. Biophys. Acta* **436**:335.

132. Mosher, D.F. (1977), *Biochim. Biophys. Acta* **491**:205.

133. Gahmberg, C.G. and Hakomori, S.-I. (1973), *Proc. Natl. Acad. Sci. USA* **70**:3329.

134. Vaheri, A. and Ruoslahti, E. (1975), *J. Exp. Med.* **142**:530.

135. Itaya, K. and Hakomori, S.-I. (1976). *FEBS Lett.* **66**:65.

136. Pearlstein, E. and Waterfield, M.D. (1974), *Biochim. Biophys. Acta* **362**:1.

137. Hynes, R.O. and Humphryes, K.C. (1974), *J. Cell Biol.* **62**:438.

138. Hynes, R.O. (1979), in *Surfaces of Normal and Malignant Cells* (R.O. Hynes, Ed.) p. 103, John Wiley & Sons, Inc., New York.

139. Robbins, P.W., Wickus, G.G., Branton, P.E., Gaffney, B.J., Hirschberg, C.B., Fuchs, P., and Blumberg, P.M. (1974), *Cold Spring Harbor Symp. Quant. Biol.* **39**:1173.

140. Hynes, R.O. and Wyke, J.A. (1975), *Virology* **64**:492.

141. Gallimore, P.H., McDougall, J.K., and Chen, L.B. (1977), *Cell* **10**:669.

142. Becker, D., Kurth, R., Critchley, D., Friis, R., and Heinz, B. (1977), *J. Virol.* **21**:1042.

143. Yamada, K.M., Ohanian, S.H., and Pastan, I. (1976), *Cell* **9**:241.

144. Yamada, K.M., Yamada, S.S., and Pastan, I. (1976), *Proc. Natl. Acad. Sci. USA* **73**:1217.

145. Willingham, M.O., Yamada, K.M., Yamada, S.S., Pouyssegur, J., and Pastan, I. (1977), *Cell* **10**:375.

146. Yamada, K.M., Olden, K., and Pastan, I. (1978), *Ann. N.Y. Acad. Sci.* **312**;256.

147. Yamada, K.M. and Pastan, I. (1976), *J. Cell Physiol.* **89**:827.

148. Hynes, R.O., Destree, A.T., Mautner, V.M., and Ali, I.U. (1977), *J. Supramol. Struct.* **7**:397.

149. Olden, K. and Yamada, K.M. (1977), *Cell* **11**:957.

150. Holley, R.W. and Kiernan, J.A. (1968), *Proc. Natl. Acad. Sci. USA* **60**:300.

151. Holley, R.W. and Kiernan, J.A. (1974), *Proc. Natl. Acad. Sci. USA* **71**:2908.

152. Stocker, M.G.P. (1973), *Nature* **246**:200.

153. Dulbecco, R. and Elkington (1973), *Nature* **246**:197.

154. Stoker, M. and Piggott, D. (1974), *Cell* **3**:207.

155. Smith, M.S., Scher, C.D., and Todaro, G.J. (1970), *Bacteriol. Proc.* **217**:187.

156. Sanders. F.K. and Burford, B.O. (1964), *Nature* **201**:786.

157. Macpherson, I. and Montagnier, L. (1964), *Virology* **23**:291.

158. Graham, J.M. (1979), in *Surfaces of Normal and Malignant Cells* (R.O. Hynes, Ed.), p. 199, John Wiley & Sons, New York.

159. Burk, R.R. (1968), *Nature* **219**:1272.

160. Anderson, W.B., Johnson, G.S., and Pastan, I. (1973), *Proc. Natl. Acad. Sci. USA* **70**:1055.

161. Peery, C.V., Johnson, G.S., and Pastan, I. (1971), *J. Biol. Chem.* **246**:5788.

162. Hatanaka, M. and Hanafusa, H. (1970), *Virology* **41**:647.

163. Weber, M.J. (1973), *J. Biol. Chem.* **248**:2978.

164. Yoshida, M. and Ikawa, Y. (1977), *Virology* **83**:444.

165. Dubrow, R., Pardee, A.A., and Pollack, R. (1978), *J. Cell Physiol.* **95**:203.

166. Royer-Pokora, B., Beug, H., Claviez, M., Winkhardt, H.J., Friss, R.R., and Graaf, T. (1978), *Cell* **13**:751.

167. Schwarz, R.I., Farson, D.A., Soo, W.J., and Bissel, M.J. (1978), *J. Cell Biol.* **79**:672.

168. Perbal, B., and Rassoulzadegan, M. (1980), *J. Virol.* **33**:697.

169. Kletzien, R.F. and Perdue, J.F. (1974), *J. Biol. Chem.* **249**:3375.

170. Reich, E. (1973), *Fed. Proc.* **32**:2174.

171. Yunis, A.A., Schultz, D.R., and Sato, G.H. (1973), *Biochem. Biophys. Res. Commun.* **52**:1003.

172. Christman, J.K. and Acs, G. (1974), *Biochim. Biophys. Acta* **340**:339.

173. Goldberg, A.R. (1974), *Cell* **2**:95.

174. Pollack, R., Risser, R., Conlon, S., and Rifkin, D. (1974), *Proc. Natl. Acad. Sci. USA* **71**:4792.

175. Laug, W.E., Jones, P.A., and Benedict, W.F. (1975), *J. Natl. Cancer Inst.* **54**:173.

176. Reich, E. (1978), in *Molecular Basis of Biological Degradative Processes* (R.D. Berlin, H. Herrmann, I.H. Lepow, and J.M. Tanzer, Eds.), p. 155, Academic Press, New York.

177. Ossowski, L., Unkeless, J.C., Tobia, A., Quigley, J.P., Rifkin, D.B., and Reich, E. (1973), *J. Exp. Med.* **137**:112.

178. Unkeless, J.C., Tobia, A., Ossowski, L., Quigley, J.P., Rifkin, D.B., and Reich, E. (1973), *J. Exp. Med.* **137**:85.

179. Unkeless, J.C., Dano, K., Kellerman, G.M., and Reich, E. (1974), *J. Biol. Chem.* **249**:4295.

180. Reich, E. (1975), in *Proteases and Biological Control* (E. Reich, D.B. Rifkin, and E. Shaw, Eds.), p. 333, Cold Spring Harbor Laboratory, Cold Spring Harbor, New York.

181. Christman, J.K., Silverstein, S.C., and Acs, G. (1977), in *Research Monographs in Cell and Tissue Physiology, Vol. 2, Proteinases in Mammalian Cells and Tissues.* (A.J. Barret, Ed), p. 90, North Holland, Amsterdam.

182. Quigley, J.P. (1979), In *Surfaces of Normal and Malignant Cells* (R.O. Hynes, Ed.), p. 247, John Wiley & Sons, New York.

183. Jones, P., Benedict, W., Strickland, S., and Reich, E. (1975), *Cell* **5**:323.

184. Coleman, P.L., Latham, H.G., Jr., and Shaw, E.N. (1976), in *Methods in Enzymology,* Vol. 45, Academic Press, New York.

185. Zimmerman, M., Quigley, J.P., Ashe, B., Dorn, C., Goldfarb, R., and Troll, W. (1978), *Proc. Natl. Acad. Sci. USA* **75**:750.

186. Dano, K. and Reich, E. (1979), *Biochim. Biophys. Acta* **566**:138.

187. Summaria, L., Arzadon, L., Bernabe, P., and Robbins, K.C. (1972), *J. Biol. Chem.* **247**:4691.

188. Dano, K. and Reich, E. (1975), in *Proteases and Biological Control* (E. Reich, D.B. Rifkin, and E. Shaw, Eds.) p. 357, Cold Spring Harbor Laboratory, Cold Spring Harbor, New York.

189. Granelli-Piperno, A. and Reich, E. (1978), *J. Exp. Med.* **147**:223.

190. Christman, J.K., Acs, G., Silagi, S., and Silverstein, S.C. (1975), in *Proteases and Biological Control* (E. Reich, D.B. Rifkin, and E. Shaw, Eds.), p. 827, Cold Spring Harbor Laboratory, Cold Spring Harbor, New York.

191. Quigley, J.P. (1979), *Cell* **17**:131.

192. Rifkin, D.B. and Pollack, R. (1977), *J. Cell Biol.* **73**:47.

193. Roblin, R.O., Young, P.L., and Bell, T.E. (1978), *Biochem. Biophys. Res. Commun.* **82**:165.

194. Yamanishi, K. and Rapp, F. (1979), *J. Virol.* **31**:415.

195. Howett, M.K., High, C.S., and Rapp, F. (1978), *Cancer Res.* **38**:1075.

196. Adelman, S.F., Howett, M.K., and Rapp, F. (1980), *J. Gen. Virol.* **5**:101.

197. Perbal, B. (1980), *J. Virol.* **35**:420.

198. Wolf, B.A. and Goldberg, A.R. (1978), *Virology* **89**:570.

199. Rifkin, D.B., Beal, L.P., and Reich, E. (1975), in *Proteases and Biological Control* (E. Reich, D.B. Rifkin, and E. Shaw, Eds.), p. 841, Cold Spring Harbor Laboratory, Cold Spring Harbor, New York.

200. Beers, W.H., Strickland, S., and Reich, E. (1975), *Cell* **6**:387.

201. Strickland, S., Reich, E., and Sherman, M.I. (1976), *Cell* **9**:231.

202. Rifkin, D.B. (1978), *J. Cell Physiol.* **97**:421.

203. Wigler, M. and Weinstein, I.B. (1976), *Nature* **259**:232.

204. Vassali, J.D., Hamilton, J., and Reich, E. (1976), *Cell* **8**:271.

205. Granelli-Piperno, A., Vassali, J.D., and Reich, E. (1977), *J. Exp. Med.* **146**:1623.

206. Strickland, S. and Beers, W.H. (1976), *J. Biol. Chem.* **251**:5694.

207. Seifert, S.C. and Gelehrter, T.D. (1978), *Proc. Natl. Acad. Sci.* **75**:6130.

208. Werb, Z. (1978), *J. Exp. Med.* **147**:1695.

209. Strickland, S. and Mahdavi, V. (1978), *Cell* **15**:393.

210. Wilson, E.L. and Reich, E. (1978), *Cell* **15**:385.

211. Lee, L.S. and Weinstein, I.B. (1978), *Nature* **274**:696.

212. Mott, D.M., Fabisch, P.H., Sani, B.P., and Sorof, S. (1974), *Biochem. Biophys. Res. Commun.* **61**:621.

213. Goldberg, A.R., Wolf, B.A., and Lefevre, P.A (1975), in *Proteases and Biological Control* (E. Reich, D.B. Rifkin, and E. Shaw, Eds.), p. 857, Cold Spring Harbor Laboratory, Cold Spring Harbor, New York.

214. Jones, P.A., Laug, W.E., and Benedict, W.F. (1975), *Cell* **6**:245.

215. Stiles, C.D., Desmond, I.W., Chuman, L.M., Sato, G., and Saier, M.H. (1976), *Cancer Res.* **36**:1353.

216. Stevens, D.A., Easty, G.C., and Ambrose, E.J. (1964), *Br. J. Cancer* **17**:719.

217. Vigier, P. (1969), *Int. J. Cancer* **4**:859.

218. Boone, C.W., Takeichi, N., Paraujpe, M., and Gilden, R. (1976), *Cancer Res.* **36**:1626.

219. Stoker, M. and Abel, P. (1963), *Cold Spring Harbor Symp. Quant. Biol.* **27**:375.

220. Risser, R. and Pollack, R. (1974), *Virology* **59**:477.

221. Freedman, V.H. and Shin, S.-I. (1974), *Cell* **3**:355.

222. Pollack, R., Risser, R., Conlon, S., Freedman, V., Shin, S.-I., and Rifkin, D.B. (1975), *Cold Spring Harbor Conf. Cell Proliferation* **2**:885.

223. Shin, S.-I., Freedman, V.H., Risser, R., and Pollack, R. (1975), *Proc. Natl. Acad. Sci. USA* **72**:4435.

224. Duff, R. and Rapp, F. (1971), *J. Virol.* **8**:469.

225. Duff, R. and Rapp, F. (1973), *J. Virol.* **12**:209.

226. Kutinova, L., Vonka, V., and Broucer, J. (1973), *J. Natl. Cancer Inst.* **50**:759.

227. Macnab, J.C.M. (1974), *J. Gen. Virol.* **24**:143.

228. Macnab, J.C.M. (1975), *Second International Symposium on Oncogenesis and Herpes Viruses* **1**:227 (IARC Publications).

229. Kimura, S., Flannery, V.L., Liez, B., and Schaffer, P.A. (1975), *Int. J. Cancer* **15**:786.

230. Boyd, A. and Orme, T.W. (1975), *Int. J. Cancer* **16**:526.

231. Boyd, A., Orme, T., and Boone, C. (1975), *Second International Symposium on Oncogenesis and Herpes Viruses* **1**:429. (IARC Publications).

232. Darai, G. and Munk, K. (1976), *Int. J. Cancer* **18**:469.

233. Darai, G., Braun, R., Flugel, R.M., and Munk, K. (1977), *Nature* **265**:744.

234. Kucera, L.S., Gusdon, J.P., Edwards, I., and Herbst, G. (1977), *J. Gen. Virol.* **35**:473.

235. Macnab, J.C.M. (1979), *J. Gen. Virol.* **43**:39.

236. Diderholm, H., Berg, R., and Wesslen, T. (1966), *Int. J. Cancer* **1**:139.

237. Diderholm, H., Stenkvist, B., Ponten, J., and Wesslen, T. (1965), *Exp. Cell Res.* **37**:452.

238. Tooze, J. (1980), in *DNA Tumor Viruses* (J. Tooze, Ed.), p. 205–235, Cold Spring Harbor Laboratory, Cold Spring Harbor, New York.

239. Albrecht, T. and Rapp, F. (1973), *Virology* **55**:53.

240. Robinson, R.A., Henry, B.E., Duff, R.G., and O'Callaghan, D.J. (1980), *Virology* **101**:335.

241. Li, L.-J., Jerkofsky, M.A., and Rapp, F. (1975), *Int. J. Cancer* **15**:190.

242. Takahashi, M. and Yaminishi, K. (1974), *Virology* **61**:306.

243. Kucera, L.A. and Gusdon, J.P. (1976), *J. Gen. Virol.* **30**:257.

244. Schroder, C.H., Kaerner, H.C., Munk, K., and Darai, G. (1977), *Intervirology* **8**:164.

245. Henle, W., Kiehl, V., Kohn, G., zur Hausen, H., and Henle, G. (1967), *Science* **157**:1064.

246. Pope, J.H. (1967), *Nature* **216**:810.

247. Gerber, P. and Monroe, J. (1968), *J. Natl. Cancer Inst.* **40**:855.

248. Pope, J.H., Horne, M.K., and Scott, W. (1968), *Int. J. Cancer* **3**:857.

249. Gerber, P., Whang-Peng, J., and Monroe, J.H. (1969), *Proc. Natl. Acad. Sci. USA* **63**:740.

250. Pope, J.M., Horne, M.K., and Scott, W. (1969), *Int. J. Cancer* **4**:255.

251. Nilsson, K. (1971), *Int. J. Cancer* **8**:432.

252. Jondal, M. and Klein, G. (1973), *J. Exp. Med.* **138**:1365.

253. Pattengale, P.K., Smith, R.W., and Gerber, P. (1973), *Lancet* **2**:93.

254. Orellana, T. and Kieff, E. (1977), *J. Virol.* **22**:321.

255. Miller, G.T., Shope, T., Lisco, H., Still, D., and Lipman, M. (1972), *Proc. Natl. Acad. Sci. USA* **69**:383.

256. Miller, G.T. and Lipman, M. (1973), *J. Exp. Med.* **138**:1398.

257. Hinuma, Y., Konn, M., Yamaguchi, J., Wudarski, D., Blakeslee, J., and Grace, J. (1967), *J. Virol.* **1**:1045–1051.

258. Menezes, J., Leibold, W., and Klein, G. (1975), *Exp. Cell Res.* **92**:478.

259. Albrecht, T. and Rapp, F. (1973), *Virology* **55**:53.

260. Geder, L., Lausch, R.N., O'Neill, F., and Rapp, F. (1976), *Science* **192**:1134.

261. Geder, L., Kreider, J., and Rapp, F. (1977), *J. Natl. Cancer Inst.* **58**:1003.

262. Fong, J. and Hsiung, L. (1973), *Proc. Soc. Exp. Med. Biol.* **144**:974.

263. Zur Hausen, H. (1980), in *DNA Tumor Viruses* (J. Toose, Ed.), pp. 747–797, Cold Spring Harbor Laboratory, Cold Spring Harbor, New York.

264. Todaro, G.J. and Green, H. (1966), *Virology* **28**:756.

265. Kimura, G., Itagaki, A., and Summers, J. (1975), *Int. J. Cancer* **15**:694.

266. Porasad, I., Zouzias, D., and Basilico, C. (1976), *J. Virol.* **18**:436.

267. Rassoulzadegan, M., Perbal, B., and Cuzin, F. (1978), *J. Virol.* **28**:1.

268. Fried, M. (1965), *Proc. Natl. Acad. Sci. USA* **53**:486.

269. DiMayorca, G., Callender, J., Marin, G., and Giordano, R. (1969), *Virology* **38**:126.

270. Eckhart, W. (1969), *Virology* **38**:120.

271. Kimura, G. and Dulbecco, R. (1973), *Virology* **52**:529.

272. Kimura, G. and Itagaki, A. (1975), *Proc. Natl. Acad. Sci. USA* **72**:673

273. Brugge, J.S. and Butel, J.S. (1975), *J. Virol.* **15**:619.

274. Kimura, G. and Itagaki, A. (1975), *Proc. Natl. Acad. Sci. USA* **72**:673.

275. Martin, R.G. and Chou, J.Y. (1975), *J. Virol.* **15**:599.

276. Osborn, M. and Weber, K. (1975), *J. Virol.* **15**:636.

277. Tegtmeyer, P. (1975), *J. Virol.* **15**:613.

278. Seif, R. and Cuzin, F. (1977), *J. Virol.* **24**:721.

279. Rassoulzadegan, M. and Cuzin, F. (1980), *J. Virol.* **33**:909.

280. Nonoyama, M. and Pagano, J. (1972), *Nature* **238**:169.

281. Adams, A., Lindahl, T., and Klein, G. (1973), *Proc. Natl. Acad. Sci. USA* **70**:2888.

282. Kawai, Y., Nonoyama, M., and Pagano, J. (1973), *J. Virol.* **12**:1006.

283. Nonoyama, M. and Pagano, J. (1973), *Nature* **242**:44.
284. Pritchett, R., Pedersen, M., Kieff, E. (1976), *Virology* **74**:227.
285. Adams, A. and Lindahl, T. (1975), *Proc. Natl. Acad. Sci. USA* **72**:1477
286. Lindahl, T., Adams, A., Bjursell, G., Bornkamm, G.W., Kaschka-Dierich, C., and Jehn, U. (1976), *J. Mol. Biol.* **102**:511.
287. Adams, A., Bjursell, G., Kaschka-Dierich, C., and Lindahl, T. (1977), *J. Virol.* **22**:373.
288. Raab-Traub, N., Pritchett, R., and Keiff, E. (1978), *J. Virol.* **27**:388.
289. Kaschka-Dierich, C., Nazerian, K., and Tomssen, R. (1979), *J. Gen. Virol.* **44**:271.
290. Werner, F.J., Bornkamm, G.W., and Fleckenstein, B. (1977), *J. Virol.* **22**:794.
291. Munyon, W., Kraiselburd, E., Davis, D., and Mann, J. (1971), *J. Virol.* **1**:813.
292. Davidson, R.L., Adelstein, S.J., and Oxman, M.N. (1973), *Proc. Natl. Acad. Sci. USA* **70**:1912.
293. Jamieson, A.T., Macnab, J.C.M., Perbal, B., and Clements, J.B. (1976), *J. Gen. Virol.* **32**:493.
294. Bacchetti, S. and Graham, F. (1977), *Proc. Natl. Acad. Sci. USA* **74**:1590.
295. Maitland, N.J. and McDougall, J.K. (197), *Cell* **11**:233.
296. Wigler, M., Silverstein, S., Lee, L., Pellicer, A., Cheng, Y., and Axel, R. (1977), *Cell* **11**:223–232.
297. Minson, A.C., Wildy, P., Buchan, A., and Darby, G. (1978), *Cell* **13**:581.
298. Munyon, W., Buchsbaum, R., Paoletti, E., Mann, J., Kraiselburd, E., and Davis, D. (1972), *Virology* **49**:683.
299. Davidson, R., Adelstein, S., and Oxman, M. (1973), *Proc. Natl. Acad. Sci. USA* **70**:1912.
300. Thouless, M. and Wildy, P. (1975), *J. Gen. Virol.* **26**:159.
301. Leiden, J., Buttyan, R., and Spear, P. (1976), *J. Virol.* **20**:413.
302. Sugino, W.M., Chadha, K.C., and Kingsbury, D. (1977), *J. Gen. Virol.* **36**:111.
303. Smiley, J.R., Steege, D.A., Juricek, D.K., Summers, W.P., and Ruddle, F.H. (1978), *Cell* **15**:455.
304. Duff, R. and Rapp, F. (1971), *Nature* **233**:48.

305. Copple, C.D. and McDougall, J.K. (1976), *Int. J. Cancer* **17**:501.

306. Galloway, D.A., Copple, C.D., and McDougall, J.K. (1980), *Proc. Natl. Acad. Sci. USA* **77**:880.

307. Camacho, A. and Spear, P.G. (1978), *Cell* **15**:993.

308. Reyes, G., LaFemina, R., and Hayward, G.S. (1979), *Cold Spring Harbor Symp. Quant. Biol.* **44**:629.

309. Collard, W., Thornton, H., and Green, M. (1973), *Nature* **243**:264.

310. Dambaugh, T., Nkrumah, F.K., Biggar, R.J., and Kieff, E. (1979), *Cell* **16**:313.

311. Hayward, S.D. and Kieff, E.D. (1976), *J. Virol.* **18**:518.

312. Powell, A.L.T., King, W., and Kieff, E.D. (1979), *J. Virol.* **29**:261.

313.. Reedman, B.M. and Klein, G. (1973), *Int. J. Cancer* **11**:499.

314. Roizman, B. and Kieff, E.D. (1975), in *Cancer: A Comprehensive Treatise* (F.F. Becker, Ed.), Vol 2, p. 241, Plenum Press, New York.

315. Roizman, B. and Spear, P.G. (1980), in *DNA Tumor Viruses,* Cold Spring Harbor Laboratory, Cold Spring Harbor, New York.

316. Noonan, C.A., Brugge, J.S., and Butel, J.S. (1976), *J. Virol.* **18**:1106.

317. Seif, R. and Martin, R.G. (1979), *J. Virol.* **32**:979.

318. Lania, L., Gandini-Attardi, D., Griffiths, M., Cooke, B., De-Cicco, D., and Fried, M. (1980), *Virology* **101**:217.

ADDENDUM

Since this book was completed, a considerable amount of progress has been made in the study of the oncogenic potential of herpes simplex virus. In their excellent review, D. Galloway and J. McDougall (1) present a summary of the different approaches undertaken to identify the HSV genes responsible for transformation. The experiments discussed indicate that the viral genome may encode at least three transforming "genes" being used differently in various selection procedures. These genes, which may have diverged between HSV1 and HSV2, are delineated on the genome by restriction fragments BgIII I (0.31 –.42 map units), BgIII N (0.58-0.63 map units), and BgIII C (0.43-0.58 map units). There is also growing evidence that cells can remain transformed even in the absence of detectable viral DNA sequences and that viral transforming proteins may not be needed. Comparison with downstream promotion induced by avian leukosis virus (2) is tempting and the important question raised is to determine whether HSV is able to induce expression of cellular oncogenes that, in turn, would be responsible for cellular transformation.

Another promising development in the herpes field emerged from studies conducted by N. Frenkel's group (Chicago) and N. Stow (Glasgow), who located an origin of replication in the repeated sequences of HSV1 DNA (3,4) and used such subsets of HSV DNA sequences to derive cloning/amplifying vectors (amplicon) that can replicate in eucaryotic cells in the presence of standard HSV helper viruses (5).

REFERENCES

1. Galloway, D. and McDougall, J. (1983), *Nature* **302**: 21–24.
2. Neel, B.J., Hayward, W.S., Robinson, H.L., Fang, J., and Astrin, S.M (1981), *Cell* **23**:323.
3. Vlazny, D.A. and Frenkel, N. (1981), *Proc. Natl. Acad. Sci. USA* **78**:742.
4. Stow, N. (1982), *The EMBO J.* **1**:863.
5. Spaete, R.R. and Frenkel, N. (1982), *Cell* **30**:295.

INDEX

LIBRARY DAVID.